BEHIND THE MASK
OF FACEBOOK

BEHIND THE MASK OF FACEBOOK:

A WHISTLEBLOWER'S SHOCKING STORY OF BIG TECH BIAS AND CENSORSHIP

Ryan Hartwig and
Kent Heckenlively, JD

Skyhorse Publishing books may be purchased in bulk at special discounts for sales promotion, corporate gifts, fund-raising, or educational purposes. Special editions can also be created to specifications. For details, contact the Special Sales Department, Skyhorse Publishing, 307 West 36th Street, 11th Floor, New York, NY 10018 or info@skyhorsepublishing.com.

Skyhorse® and Skyhorse Publishing® are registered trademarks of Skyhorse Publishing, Inc.®, a Delaware corporation.

Visit our website at www.skyhorsepublishing.com.

10 9 8 7 6 5 4 3 2 1

Library of Congress Cataloging-in-Publication Data is available on file.

Cover design by Carissa Eich

Print ISBN: 978-1-5107-6794-2
Ebook ISBN: 978-1-5107-6795-9

Printed in the United States of America

"There is no swifter route to the corruption of thought
than the corruption of language."

—George Orwell

This book is dedicated to the millions of Americans whose voices were silenced by Big Tech during the 2020 US election. It is also dedicated to the billions of people throughout the world who feel the oppressive influence of Big Tech in their respective countries.

Contents

Foreword

As a former business executive in leadership positions in various Fortune 500 companies for the last twenty-five years, I have come to appreciate former employees who have blown the whistle and exposed fraud and corruption, sometimes at great personal expense.

After learning about Ryan Hartwig and his decision to go public, I recognized in Ryan some of those same attributes I've seen in other whistleblowers: courage, resilience, and honor.

Ryan was, for all intents and purposes, a low-level employee working for Cognizant, contracted to do Facebook's content moderation. But the magnitude of what he uncovered will have repercussions for Big Tech and the censorship debate for years to come.

What I appreciate about Ryan's book is how he sticks with the facts, and his thorough analysis leaves little to speculation. You'll see within this book, in black and white, Facebook's own decisions with regard to content. Whether it's protecting Greta Thunberg or making newsworthy exceptions to allow Don Lemon's hate speech, it's clear Facebook broke their own rules, time and time again. We no longer have to rely on scripted answers during Senate hearings where Big Tech executives say they'll get back to us. We already know what Facebook did because it's written in this book. Facebook censored political speech and targeted conservative speech.

Ten years from now, it could be leftist speech that is being singled out, which is why this issue is so important. We need to protect free speech for everyone. Facebook has incredible power, and this influence has gone unchecked, gradually consuming more and more control. They have become a de facto government and can exert their influence within any aspect of our lives, including elections and healthcare.

Ryan delves into these issues, asking important questions about why Facebook has assumed the authority to monitor our elections. He also asks the same question about foreign elections and the amount of influence Facebook has internationally. Facebook can single-handedly influence an election or allow its tools to be manipulated by dictators and AI bots intent on crushing the political competition through propaganda and smears.

Regardless of your political persuasion, this book takes us deep into the inner sanctum of Facebook's policy decisions. We all know Facebook manipulates its users to spend more and more time on the platform, and Ryan's insights show us how Facebook was monitoring ideological gaps between users and segmenting users according to their political views.

Facebook's clear failures with regard to privacy are not the only area where they have demonstrated their complete inability to protect users. They have also failed in the content moderation business and targeted users because of their political viewpoints. By spending billions of dollars on content moderation and using their AI to censor content, Facebook has become a threat to our American republic. And not surprisingly, many of our elected representatives are complicit in ceding control of the largest public cyberspace to Mark Zuckerberg, a few policy advisors, and a hand-picked oversight board that has zero interest in aligning itself with American values or respecting the rule of law.

I urge you to read this book to learn just how corrupt Facebook has become. You will learn how Facebook made exceptions to allow child pornography, child abuse, and hate speech all in the name of political correctness. Facebook censored millions of users and viral videos in order to promote a certain agenda. It's time to stop Facebook in its tracks, and by educating ourselves on their inner workings, we can more effectively stand against the most powerful censorship tool of the twenty-first century.

—Patrick Byrne,
founder and former CEO of overstock.com

Prologue:
An Unusual Job Interview

The interviewer slid his laptop over and said, "I want you to flip that around and scroll through these images. Let me know if you're comfortable viewing them."

I clicked through the pictures. There was a headless body, anime porn, nudity, and some very graphic porn.

"Can I skip any of these?"

"Not if you want the job," Carlos* replied.

This might sound like the world's strangest job interview. But I was prepared for anything. In January 2018, I'd been working for four months as a security guard for Eagle-Force Security. I'd been assigned to a furniture store called Conn's Plus in Phoenix near the I-17 freeway and Dunlap. It wasn't the best part of town. My job was to keep homeless people and young toughs from wandering into the store. You should know that like a lot of people in my generation, I've worked in a number of jobs, many of them in security.

One day at work, two guys walked into the furniture store looking to buy some headphones. I had a casual conversation with them. Somehow we got talking about what they do for a living. They said they worked for Cognizant right on the other side of the freeway. They explained the company had a brand-new project and was hiring.

In a roundabout way, they hinted at the kind of work it was—reviewing social media for "the client"—but they had to be careful about how they worded it. However, based on their description, the idea intrigued me. I especially liked the fact there was a Spanish department where I could put

---

* A pseudonym.

my bilingual skills to good use. So, I applied, which is why I was now look-ing at graphic porn and headless bodies during a job interview.

The first part of the interview had been more traditional. I did well in the Spanish language portion of the exam, which was surprising to them considering I'm a pale white guy with blond hair and blue eyes. Growing up in Arizona, I'd been fascinated by the Mexican culture, taken Spanish classes in school, and as a Mormon[†] done my missionary work in the coun-try. As a result, I could write like a well-educated Mexican professional, in addition to being fluent in conversational Spanish.

Then came the images.

Considering what came after, maybe I should have declined to view the images. But there was something that struck at my moral core. I knew the cartels were using social media to broadcast their brutality and terror-ize communities. And I understood that young people were the most avid consumers of social media. I didn't think porn should be on the platforms, corrupting their burgeoning ideas of sexuality.

But my training as a security guard kicked in. We don't run away from trouble; we run toward it. We do what needs to be done, without fear or favoritism.

"Yeah, I can deal with this," I replied.

When I started the job at Cognizant, they told me the name of the client.

Facebook.

I started the job believing I'd be guarding the platform from the cartels and pornographers.

Instead, Facebook would declare that a large part of the American pub-lic was the real enemy.

---

† Members of the Church of Jesus Christ of Latter-Day Saints are known as Mormons.

CHAPTER ONE

Getting the Job and
Leaving the Nest

After talking to the two guys in the furniture store, I went to the Cognizant website and applied for the job of "Bilingual Social Media Content Moderator." After filling in all the relevant information, I took an on-site cognitive e test. It seemed similar to a mini-SAT test one might take in high school, giving you passages to read and then asking questions about the text. There were also a number of questions about computers and social media, such as the different kinds of browsers, keyboard shortcuts, and common computer acronyms like "What does LOL mean?"

I finished the cognitive test and submitted it. I left a few questions unanswered, and that worried me. But a week later they emailed me to set up an in-person interview.

I dressed in a suit and tie for the interview and went to the Cognizant offices, located on the top floor of a nondescript six-story office building just off the I-17 freeway in north-central Phoenix. Next door to the office building was a Sheraton Hotel.

In the main lobby, I pushed the call button, gave them my name, and said I had an interview at Cognizant. They sent the elevator down for me.

When the doors opened on the sixth floor, it was as if I'd boarded the elevator in Phoenix and was now in Silicon Valley. There was a large, whimsical cartoonish mural on one entire wall. It showed the sun rising in the East over the mountains surrounding Phoenix, some tall buildings, a road through a desert with cactuses and palm trees, birds flying in the air, a rabbit sleeping by the side of the road, and a wild peccary in the distance. The people were in balance with nature, perfect combinations of human and animal spirits. There was an older guy in shorts and a belly looking at his phone, but he had a deer's head, complete with antlers. There was a stylish woman riding her bike, and she had the head of a blue fox. Near the right corner of the mural was a guy hiking through the desert, his bird's head looking at the city and giving a thumbs-up, as if to say, "There's nothing we can't do." Above the birdman, a hot-air balloon floated in the sky. The colors were bold, vibrant, and cheerful. I immediately liked the vibe of the place.

I was interviewed by three people, Carlos, Tim, and Jane.* In the first part of the interview, they reviewed my résumé, and we just engaged in general conversation. They noted with approval I had my bachelor's degree from Arizona State University. Most of the hires did not have college degrees. Some were fresh out of high school, and others were in their early twenties. At age thirty, I would be one of the older hires.

In the second part of the interview, Carlos switched to Spanish and was surprised by my fluency. In addition to studying Spanish in school, I'd been posted to Mexico for two years as a missionary. In addition, my wife, Livy, is from Honduras, having come to the United States in 2005. Normally, we speak English at home, although we do often converse in Spanish. Unlike many of the people hired by Cognizant who spoke Spanish at home with their parents, I also had academic training. My Spanish fluency, both for conversation and academic writing, was comparable to a well-educated Venezuelan or Colombian. I met many of these Venezuelans and Colombians when I worked in the Spanish language division at Cognizant, and there seemed to be an easy and natural sympathy between us.

It was after this discussion that Carlos passed his laptop over to me and asked me to flip it around to look at some pictures. The images gave me pause, but I wanted the job. It was clear the company was just getting started, and I wanted to be on the ground floor. I knew this would be the final part of the job interview, whether I could handle these images.

I told them I was fine with the images, and we finished up a few minutes later. They thanked me for coming in, and I left.

* * *

---
* All pseudonyms.

About a week later, I got the call from Cognizant that I'd been hired. I needed to give two weeks' notice at my security guard job, and so my first day of work for Cognizant was March 21, 2018.

Although the pay wasn't great, a little over twenty-eight thousand dollars a year, you received health benefits from the first day of work. That was a big selling point with my wife. In addition, the office was just about a seven-minute drive from our apartment. My wife had been working as a dental assistant for several years but was interested in working with children. With the health benefits in place, she felt more comfortable about looking to make a career switch.

For the first few days, I was simply dealing with a number of human resources forms and policies, such as the fact that you couldn't bring your cell phone into the workplace. It was stored in a locker outside the work area. If you were caught with your cell phone on you, it could result in your being immediately fired.

The training class for my cohort began on March 21, 2018, without about thirty of us in the Spanish language group. I was the only white guy in my group, but that didn't matter. I've always loved the warmth and friendliness of the Hispanic culture, and when it was clear I was fluent in Spanish, I got a lot of respect with comments like "El gringo habla bien el español." The two trainers for the class were Shawn Browder and Veronica Castillo†, names that would become important, as they'd be shown in the Project Veritas videos, which would be released in 2020.

In the month of training, we went through approximately thirteen different Cognizant policies, such as the Bullying Policy or Graphic Violence, and we'd have a quiz about every two days to make sure we were keeping up with the material. I was taking notes constantly during the class and emailing them to myself at the end of the day because there was so much coming at you. In those first couple of days, they also showed us graphic pictures of pornography, violence, and torture and then explained the underlying logical principles on how to classify and deal with them.

Several members of the class started asking hypotheticals to our trainer, Veronica, and she'd reply by saying that she didn't deal with hypotheticals. You had to provide her with an actual image, and then she'd tell you how to analyze it. Each image we came across was so different in how the policy could apply to it. In addition, we also learned

† A pseudonym.

how Cognizant came to be working for Facebook, as later detailed in a 2019 article:

> On May 3, 2017, Mark Zuckerberg announced the expansion of Facebook's "community operations" team. The new employees, who would be added to 4,500 existing moderators, would be responsible for reviewing every piece of content reported for violating the company's community standards. By the end of 2018, in response to criticism of the presence of violent and exploitative content on the social network, Facebook had more than 30,000 employees working on safety and security, about half of whom were content moderators.
>
> The moderators include some full-time employees, but Facebook relies heavily on contract labor to do the job . . .
>
> The use of contract labor also has a practical benefit for Facebook: it is radically cheaper. The median Facebook employee earns $240,000 annually in salary, bonuses, and stock options. A content moderator for Cognizant in Arizona, on the other hand, will earn just $28,800 per year.[1]

It was clear to me from the start that Cognizant and Facebook were trying to accomplish something that had never been done before, namely, to come up with reasonable rules for social media. We at Cognizant helped shape these rules by giving feedback to Facebook. And at the beginning it seemed to make sense, even though it was clear they were trying to find their way. As I understood it, the office had just been set up in the fall of 2017 with about ten people. Eventually there would be more than a thousand employees working at the office, in three different shifts. Although we were all told that "the client" was Facebook, they encouraged us not to share that information with family and friends.

There was also a four-part process for determining whether content violated a policy. The first place you'd look was the policy itself, an approximately thirty-thousand-word document split up into various webpages. The second place you'd look was a document called "Known Questions," which contained issues that had already arisen and had been addressed. Then there was another document called "Operational Guidelines," which discussed guiding principles. These two documents were abbreviated as KQ and OG, respectively. The last place you'd go was the "Workplace," which tried to be a clearing house for all these issues and was where they posted one-off exceptions to the policy. A content moderator would often simply type in a word and hit Control F on their keyboard to find out where that word might be in the various documents.

As the class continued, a few people dropped out because of the violent imagery, but that was probably no more than 10 percent of the group. And as our quizzes started getting graded, I was usually near the top of the group, finishing fourth or fifth out of our group of twenty-five by the end. We were encouraged to ask a lot of questions, and I found myself stimulated by the mental challenge of it.

For example, if there's a video of two adults cursing at each other, does it get removed? The answer is no, it doesn't. But what if it's two minors cursing at each other? That does get removed. However, what if you can't tell whether the people cursing at each other are over the age of eighteen? That's a problem, right? Well, the Operational Guidelines tell you that if you can't tell, you assume they're minors and delete. That made a good deal of sense to me.

Another guiding principle we were taught was that the burden of clarity is on the user. If there's something that's ambiguous, we were told to take it down. If it seems wrong, but we're not sure, we should delete it. If we don't see a problem, then it's fine. But if there's a possible violation and the user wasn't clear about their intent, it gets deleted because they should have made sure we understood the meaning.

I've mentioned my Mormon background, but can't say I'm exceptionally devout. I believe in God, but for many years I didn't attend services. My wife wasn't Mormon when I married her, but shortly after, she converted, and we started attending services more regularly. But I'd say I'm culturally Mormon, and that pertains to my use of profanity.[‡] If you look at the website for the Church of Jesus Christ of Latter-Day Saints (referred to as the Mormon church), this is what you'd find as their official position on "bad" language:

> Profanity is disrespect or contempt for sacred things. It includes casual or irreverent use of the name of any member of the Godhead. It also includes any type of unclean or vulgar speech or behavior.
>
> We should always use the names of Heavenly Father, Jesus Christ, and the Holy Ghost with reverence and respect. Misusing their names is a sin. Profane, vulgar, or crude language or gestures, as well as immoral jokes, are offensive to the Lord and to others.
>
> Foul language is both degrading and harmful to the spirit. We should not let others influence us to use foul language. Instead, we should use clean

---

‡ Although I adhered to the profanity guidelines of the church for the better part of my life, I now do swear occasionally.

language that uplifts and edifies others, and we should choose friends who use good language. If friends and acquaintances use profanity, we can good-naturedly encourage them to choose other words. If they persist, we can politely walk away or change the subject.[2]

Does that sound like what a Mormon might say? Do you laugh because it makes us sound squeaky clean? I don't mind telling you it resonates with me and is how I was raised to behave. Like many Mormons, I would describe myself as politically conservative, and I like to talk about politics. But I always conduct those conversations with respect.

At the beginning, Cognizant preached a similar philosophy that was extremely compatible with the Mormon ideal. From the start of our training, they instructed us to treat one another with DKR (dignity, kindness, and respect) because we were going to be seeing some pretty awful things and it was important we be good to one another. In keeping with that principle, during our training we received a visit from the staff psychiatrist, who told us we were going to be seeing a lot of disturbing things and that he and the other counselors were always on call if we needed to discuss an issue.

If there was anything that struck me as slightly odd during the training, it was that one of our trainers, Veronica Castillo, said that Barack Obama was her Patronus Charm. The Patronus Charm comes from the Harry Potter series of novels and is one of the most powerful charms, invoking an individual's spirit guardian. In essence, she was saying that Barack Obama was her spirit guardian. Later, I'd find out that Veronica also worked for Democratic Congresswoman Gabrielle Giffords, who was shot by a crazed gunman at a campaign event in January 2011. She made a full recovery, and her husband, Mark Kelly, a former space shuttle commander, became the junior United States senator from Arizona in December 2020.

As we neared the end of our month-long training class, a sense of excitement swept through the group. We wouldn't go directly to the floor but would spend an additional two weeks doing something called "nesting."

In "nesting," we'd spend half our day reviewing and deciding whether actual posts would be deleted. The other half of the day, we'd be shadowing a working content moderator, who would tell us why they were making certain decisions and invite us to ask any questions.

* * *

I can still remember the thrill of that first Monday morning when we sat at the end of a U-shaped floor plan, just outside the training room, but in view of the production floor. We would be "actioning" content, as they called it, and our work would be scored.

We were told that our scores wouldn't count against us during this time. But when we left the training phrase, the scores would determine promotions and continued employment. They also told us not to worry about our speed at that point. Just focus on making good decisions, they told us. Eventually they'd want our average handle time (AHT) to be thirty seconds for each "job." That was the ultimate goal.

In the nesting phase, they told us how to manage our browser, because you might need to have ten different tabs open at the same time. Since we were handling the Latin American market, we often didn't know who were celebrities and who were just regular people. You had to know current events in Latin America, the personalities and the politicians, so in addition to being a content moderator, you were also something of a researcher.

If you did exceptionally well during the first three to six months, you could apply to become a subject matter expert (SME), which I wanted to be. However, you had to score a 98 percent on your jobs over the previous two-week period to be considered for a subject matter expert position, and I only scored a 97 percent. I was upset with myself because there was one job where I'd made the wrong call, and that prevented me from becoming a subject matter expert.

While during this time it was permissible to share images we might come across on our screen to the other members of our cohort, we were sternly warned not to do so with any images of child pornography. The rules were quite clear. We could ask for guidance from a supervisor, but sharing such images with the members of our cohort was a violation of federal law.

Our group seemed to have come together well through training and nesting. I remember asking lots of questions of the content moderators when I shadowed them. It was exciting to spend half the day watching experienced content moderators on the production floor make their decisions.

As the training and nesting phase wound down, I believe most of us were thinking, *Yeah, I can do this job.*

Little did I know the greatest threat came not from the images that would flood our screens from the outside world, but from intolerant and hypocritical policies that I can only assume came from "the client."

CHAPTER TWO

Finding My Groove as a Content Moderator

Any new job requires a significant amount of new learning. One could say this was especially true with content moderation at Cognizant because we were essentially building the plane as we were flying it.

Before I delve too deeply into my criticisms, I want to be fair and acknowledge some of the unique problems created by social media. It's one thing to call social media platforms the new digital town square, but there are some significant differences. If I stand up in the traditional town square, or at my local city council meeting, to express my point of view, I'm not just spouting off whatever pops into my head. If I want to be persuasive, I'm observing how the audience is reacting to what I'm saying, and I'm figuring out how I need to tailor my message to be most effective. There's a feedback loop built into traditional methods of free speech that determines whether I will gather supporters for my point of view or simply be dismissed as a crackpot.

One of the major flaws of social media is that the best arguments don't necessarily get promoted, but rather those that generate the most outrage. The crazies get the lion's share of attention, while the rational people who try to understand both sides of an issue get ignored. The goal of the social media platform is to get the most eyeballs possible on a post so they can then sell advertising to companies. Outrage is the fuel of social media, and it's likely to cause an explosion.

Let me explain to you what it was like trying to fly this plane as we were in the midst of building it.

* * *

The goal was for content moderators to eventually make decisions on at least three hundred posts a day and eventually be able to make our decisions within thirty seconds of its popping up on our screen. That meant in a week we'd make decisions on fifteen hundred posts. In a typical year, we were expected to review seventy-five-thousand pieces of content. Every week, Quality Assurance would pick about fifty random posts upon which we'd made a decision and review our work. The expectation was we'd be above 95 percent accuracy, which meant if they believed you made the wrong call on three of those posts, you were in trouble, even though that might be 94 percent accuracy. The margin for error was extremely small.

You could dispute Quality Assurance by doing something called "pushing back for points," but then the time you spent doing that would impact your productivity. Sometimes the post might go back to that same reviewer, and other times it might go to a different one. I disputed several of the calls, showing where my decision was in accord with policy, and many of those times I got my points back. While working in Quality Assurance was superior to being a content moderator, they only earned a dollar an hour more than we made.

When I started working on the production floor on the Spanish side of Cognizant, I was one of only two white guys. I was eventually given the nickname "güero," which in Spanish denotes someone with blond hair and usually means someone with light, fair skin. The Latin American culture is generally fond of nicknames for people, and I always accepted the name as a sign of affection.

The first couple months were exceptionally challenging because you were constantly trying to familiarize yourself with "the policy." Much of it was written in a highly legal style, with general rules, then exceptions, and after that there were "guiding principles." Every two weeks there would be an official update to the policies, and they'd bring us all together and sit us around a large television that played a video highlighting and explaining the changes.

I think I'm good at reading and analyzing things. But I admit I struggled with a lot of the legalese, nuances, and caveats that we were given. One of the things we learned early on was to memorize the list of Spanish slurs, because when we came across those words on our screen, they would immediately get deleted. However, that gets complicated because the same slang word in different Latin American countries can have a very different meaning. One of the common Latin American slang words is *puto*, which is essentially a curse word but can also be a homophobic slur when used as a noun. Facebook users would try all sorts of unique ways to keep it from being deleted, like spelling it out "P-U-T-O," in the hopes that we would miss it.

One of the things I think I should point out at the beginning is that the Hispanic culture tends to be family-oriented and conservative. Although in one way I might not have seemed to fit in because I'm a light-skinned white guy, in another I felt completely at home. We shared a lot of the same values.

In June 2018 we had Shawn Browder, who would later become policy manager for Cognizant but who at the time was a member of the policy team, come to give us a presentation on "Pride Month," highlighting the LGBTQ (Lesbian, Gay, Bisexual, Transgender, and Queer) community. Shawn started out by saying that Pride Month usually resulted in an increase in hate speech against that community, and we should be especially vigilant against it. The next thing he wanted to address was the fact there were going to be marches, and since some of the participants often wanted to show their pride by parading around naked, or half-naked, there was going to be an exception to the nudity policy. If a woman was protesting topless and her nipple was exposed, we were to allow it on the platform. In all other instances, a woman's nipple was deleted. But in the context of marching, it was allowed.

Many of us considered that odd, but nothing prepared me for what Browder sent in a follow-up email. I've mentioned the Hispanic culture tends to be conservative, and I think that's why he didn't want to bring it up in front of the group. One of the things we immediately deleted for hate speech was if anybody was called "filth." Makes sense, right? How can that be anything other than a slur? I was completely on board with deleting any anti-LGBTQ hate speech. Seriously, what's gained by insulting people? However, during Pride Month, we were told to make an exception if a member of the LGBTQ community called straight white males "filth," if a straight white male was not supporting LGBTQ rights. For example, the phrase "straight white males are filth for not supporting LGBT rights" would stay up and would not be deleted. That was one of the first inklings to me of the political leanings of the company, and that they weren't interested in a level playing field but wanted to tilt the discussion in a certain way.

In those first three to four months, my performance caught the attention of my supervisors, and they told me they wanted to promote me to either Quality Assurance or being a subject matter expert (SME). I knew that would only earn me an extra dollar an hour, but I still viewed us as a start-up and hoped to rise with the company. I took the test to advance from being a content moderator, but when they reviewed my QA score from the previous weeks, I had missed two questions instead of the one mistake that was allowed. I felt the decision was unfair but knew it wasn't directed at me personally.

We were hiring a lot of people during the late spring and early summer of 2018, and you could feel the excitement of being part of a company that seemed to have a great future. A good deal of our time was being taken up with monitoring the presidential elections in Mexico. I won't delve deeply into Mexican politics, other than to say they are notoriously corrupt, and in that election, a lot of hope was placed in an anticorruption candidate, Andrés Manuel López Obrador, who was often referred to simply by the acronym AMLO.

This may seem like a small example, and yet in looking back at what eventually transpired, I believe it heralded much of what was to come. Obrador's son at that time was about eleven years old, and his dark hair had streaks of blond. There's a common dessert in Mexico called chocolate flan, or chocoflan, which is a mixture of chocolate and custard. Because of his son's dark hair streaked with blond, he was often being referred to as "chocoflan," or "chocolate flan." We were told to delete all references to Obrador's son being called this nickname. At the time, it seemed like an edge case. Children of politicians didn't choose to be born into that life, so we should allow them their privacy. And yet, I couldn't help but wonder if Obrador's left-leaning politics were making Facebook put their thumb on the scales of the political debate in Mexico.

I've mentioned before that it felt like Cognizant was trying to bring a bit of that Silicon Valley culture to Arizona. In those heady early days, it was easy to believe that despite the low pay, this might end up being an outpost of that wealthy culture. We just needed to prove our value, and maybe the low pay would eventually take care of itself. For the first couple of months, there were unlimited snacks in the break room, which you could pick up on your breaks, or during the ten minutes for "wellness" you were given every day in your eight-and-a-half-hour shift. I learned later that the weekly snack bill eventually exceeded ten thousand dollars a week. After about six months, we were given a tiny key fob to log our snack usage and limited to twenty dollars a week. It should've come as no surprise that the glory days of unlimited snacks at Cognizant eventually came to an end.

When you came to work, you'd immediately go to the locker room, where you'd store your phone. We were told if you were caught on the production floor with your phone, you'd be instantly terminated. It was a strict policy but was explained to us as being necessary, since we were accessing people's private information. However, one of the many jobs I had before Cognizant was a job at the Vanguard Group processing checks, and even they didn't have such a draconian policy, even though we were accessing people's financial information.

In those first few months, I was also trying to absorb the corporate culture, the way an athlete learns a sport, so that your movements become almost automatic. One of the principles we learned was that the burden of clarity for what a post meant was on the user. That was a great, cognitive shortcut. It allowed you to make your decisions more quickly, because you didn't have to second-guess yourself. And we were encouraged to practice "self-care" and make sure we weren't being overwhelmed by the things that popped up on our screens.

Because we did see disturbing images of people behaving in monstrous ways to one another. The cartel violence was especially difficult to process. They were using Facebook to terrify their populations. I saw several videos of the cartel forcing people to their knees, then putting a shotgun up to their head and pulling the trigger. Despite my religious faith, I couldn't help but feel a complete absence of God in those instances, and the malevolent hand of evil working in human affairs. We learned about a cartel policy called "cleaning," in which the cartel would roll into town and proclaim a curfew on the citizens. They'd put up posters proclaiming nobody could be out after nine at night. Next to the poster would often be a person hanging from a rope. Sometimes they were simply hung, other times shot and hung, and the grisliest images were of people they'd burned before hanging. The cartel would take video of these signs, and the dead person hanging next to it, and post the video on Facebook.

Another cartel favorite was taking video of their execution of federal police officers they'd captured. Was there any clearer demonstration of who was in control in Mexico than the fact that these cartels could so casually kill these guardians of law and order?

I understood how these images might psychologically damage me and was frequently going to consult with the counselors. I also believed I was making a public statement to my fellow coworkers that it was okay to ask for help. The production floor was a large and open room, with somewhere between forty to fifty desks and workstations, so when I stood from my desk and walked to the counseling office, everybody could see me.

The Hispanic culture tends to be macho, not comfortable admitting vulnerability and weakness. I didn't think that strategy would serve my coworkers well and was trying to set a counterexample. Even the "gringo" needed help dealing with these images, they might say to themselves as they saw me walking by on the way to the counseling office.

One of the most disturbing videos I saw was of a man the cartel had captured. They had cut off his genitalia, and where his genitalia had been, they had a dog licking him as he screamed. It was a perversion of normal life on so many levels, a man being physically emasculated, man's best friend perversely licking him, and the screams for help that everybody who viewed it knew would go unanswered.

The counselors advised us that in order to deal with these images, we had to psychologically distance ourselves from them and not empathize. I was told to visualize myself as sitting in a movie theater, watching these images on a screen and reacting to them in a regular fashion. Then I was told to visualize a second self, standing in the back of the theater, watching my first self as I viewed and reacted to the images.

I used this strategy a great deal, especially as I viewed videos of a common cartel strategy of cutting the limbs off their prisoners with small, dull knives, as the person screamed in agony. In retrospect, I don't know how I managed to escape without significant psychological damage. But perhaps this is not so different from what soldiers see in war, or police officers, or crisis counselors, confronting the worst of human behavior, in the hope of creating a brighter future for people. That's why I thought it was important for me to use the counseling services as much as I could. Sometimes when my shift was over, I'd go home and play video games for a few hours to try and restore some semblance of normalcy to my world.

I can't say I discussed my work in depth with my wife, but I often did share some of what I saw, so she'd understand what I was experiencing.

* * *

On the Spanish language side, we also had to deal with a lot of advertisements for prostitution. A typical ad might show an attractive woman with cleavage showing, then text with some variation of "Are you looking for a good time?" and contact information. If we had those three elements—a sexually suggestive picture, some of the designated code words often used for prostitution, and contact information—we could delete. However, if it was missing one of those three elements, we couldn't delete. It had to be very clear. Therefore, the same picture, with the words "Hey, let's meet up!" or "Looking for friends!" and contact information, couldn't be deleted. Sometimes they would be explicitly direct, and you couldn't help but laugh at the brazenness of it. "Hey, I'm a prostitute and I'm looking for sex!" Or a dollar sign next to the words *Looking for sex?* Nothing like truth in advertising to make a content moderator's life easier. You can't say those users failed to meet their "burden of clarity."

Another thing that was common in the Latin American market was a woman accusing another woman of sleeping with her husband. In addition, the woman would usually accuse the other woman of being a whore or a prostitute. The policy team at Facebook made the decision that any claim of people having sex, or comments about their sexual practices, would be automatically deleted. Therefore, if Jill said, "Suzie, you slept with my husband," the comment would be deleted as a matter of policy. By the same token, if Jill said, "Suzie, you're such a whore," it would also be deleted.

However, if an allegation of a nonsexual crime was made, that would mean that the person about whom the allegation was made would have to report it to Facebook before we could take action on it. We called that a name/face match. Therefore, if Jill said, "Suzie, you stole my rings and you're an asshole," then it would be necessary for Suzie to contact us, asking for it to be taken down. The system wasn't perfect. But we'd look to see who made the complaint, and if it appeared to be the person involved, we'd take it down. However, if somebody named Bill contacted us and said, "Hey, I'm Suzie's husband and she's not an asshole," we'd have to leave it up.

At Cognizant, we had a fairly even mix of males and females. I was aware that several people left after a few weeks because they couldn't handle the images they saw. Most of the people responded in predictable ways to the images, but there was one guy who unnerved me because he'd always laugh at images of cartel violence. I don't know if that was a coping mechanism or he was some twisted freak, but I did my best to keep my distance from him.

Several cartel videos presented content moderation challenges because it wasn't enough to simply delete them. You had to delete them for the correct reason, or else Quality Assurance would mark you down for that decision. For example, say there was a thirty-second cartel video of a female prisoner. They force her to strip naked and go down to her knees, then put a gun to her head and shoot her. You might think that you'd delete that video for "Graphic Violence," right? However, in the tier system of content moderation established by Facebook (also known as the hierarchy of actions), "Nudity" was above "Graphic Violence." You'd need to watch the first fifteen seconds of the video, and if you could see an exposed nipple, or pubic hair, you'd delete the video for violating the "Nudity" policy rather than "Graphic Violence." We would also scan the thumbnails of the video for other possible policy violations.

Even the "Graphic Violence" policy had nuances to it. For example, if a person was simply shot or killed, and the camera was at some distance, you'd need to leave it up. But if you could see "Visible Innards," such as when the cartel cut off somebody's limb, and you could identify tendon or bone, you could delete it. Therefore, let's go back to our naked, kneeling woman who's shot in the head by the cartel. Maybe there's no clear image of an exposed nipple or pubic hair, so when she's shot in the head, I'm looking for any exposed brain matter.

If I don't see any, then even though it's gruesome, if there is no clear nudity or visible innards, I cannot delete it.

Yes, we were truly building the plane as we were flying it, and the decisions made were often difficult to justify.

* * *

But it wasn't always about cartel violence and sex.

We were moderating content from all Latin America, and a good deal of it was political. For example, there was content from Nicaragua, where President Daniel Ortega was perpetrating violence against his people and there were subsequent tumultuous riots in the streets. How much of that violence would we be allowed to show?

Another one of the challenges was how to cover the Venezuelan migrant crisis, which was caused by the Socialist policies of Hugo Chavez and then, after he died, his hand-picked replacement, Nicolas Maduro. Many Venezuelans had found their way to Peru, and this was causing a great deal of unrest. Many Peruvians were referring to these Venezuelan migrants as "*Venacos*" and other slurs, which we would delete en masse.

When we had a question about the meaning of a word, we were instructed to go to "Google Translate" to answer our question. Google Translate was often sufficient, but there were several times it gave me a wrong or inaccurate translation. Unlike English, which has no "standard" dictionary, there is a "standard" dictionary for the Spanish language, which is called the *Diccionario de la lengua española*. This dictionary has been in existence since the 1700s, is constantly updated as the language evolves, and is online. In addition to academic Spanish, this dictionary has slang definitions, which are further divided into the meanings of those words in different countries. I suggested to my supervisors that this was a better resource than Google Translate, but they told me I had to stick with Google.

And in our interactions with Facebook there was some good give and take. Since we were on the front line of content moderation, we saw the ways users were trying to avoid using the typical code words that would be deleted for prostitution, or creative credit-card scams. We provided that information to Facebook, and they did use it to reformulate their policies, although it seemed like it took far too long for them to make the changes. In early 2019, I remember going through some policy updates from Facebook and thinking to myself, "Didn't we tell Facebook to make these changes six months ago?"

In 2018, we were also dealing with migrant caravans from Honduras, which were being allowed to travel all through Mexico, on their way to the United States. You should realize that Mexico has extremely strict immigration laws about who can enter their country, and violation of these laws often results in a long prison sentence. But Mexico was suspending this policy for migrants traveling through their country to the United States. As content moderators, we would constantly see posts and news on Facebook with images of these caravans as they traversed through Mexico.

Another policy we had to implement was to take action on images or messages deemed "Cruel and Insensitive." For example, there was a volcano in Guatemala, the Fuego volcano, which erupted on June 3, 2018, and killed more than two hundred people. If a meme showed real pictures of the actual victims visibly suffering the tragedy, that was deleted under the policy. However, if it wasn't an actual image of a victim, perhaps a cartoon or caricature of the victim, it would simply be labeled "Mark as Cruel and Insensitive" in order to limit its reach on the platform. A similar strategy was in place regarding the fatal helicopter crash of NBA superstar Kobe Bryant. As I recall, there were no publicly available pictures of Kobe as he was dying, but if there had been, we would have deleted them, as well. As it was, there were several memes about the crash, and we marked those as "Cruel and Insensitive." So the Kobe memes wouldn't be deleted entirely, but we would limit the reach. A few other common memes we came across involved mocking victims of the Boston Marathon Bombing as well as jokes about the suicide of Robin Williams. These were all dealt with under the same "Cruel and Insensitive" policy.

But it also mattered what kind of public figure you were for content moderation at Cognizant. During this time, there was a rapper named "XXXTentacion," who was killed in a robbery on June 18, 2018. Since he was a public figure, we would normally have been told to delete any images of his death, or memes that might have been made from such pictures. However, in addition to being a public figure, he also had a prison record. Thus, in addition to being a public figure, he was also labeled a "criminal." As a result, he was exempted from the policy, and memes could be made using actual pictures of his dead body.

Despite many of the challenges, as the months flew, by I gradually began to see the fun in several aspects of my job. You got to view breaking news events and do research on public figures and events in various countries so you could make an informed decision. There was the very real problem of violence by the cartels, pornography, prostitution, and human degradation that we wanted to shut down. And in some small way, for the more common human problems, where people might bicker, miscommunicate, and call one another the types of names they'd never do if they were face to face, we'd encourage them to be a little more respectful of one another.

CHAPTER THREE

A Day in the Life of a
Content Moderator

Imagine me at Cognizant on a typical Tuesday morning in late 2018. I'd already been working for about three hours, my decision speed getting faster and faster, nearly approaching that thirty-second Holy Grail they wanted us to eventually achieve. Three hundred decisions a day, fifteen hundred a week, and seventy-five thousand a year. *Come on, Ryan, you can do it. You're doing the job of keeping the Internet safe from the cartels, pornographers, and people just being jerks to one another.*

Don't forget to mark it if you see cleavage in the image. Don't forget that an erection needs three elements, and a curved penis is not an erect penis. I clicked through a bunch of porn, easy decisions, and then came the cartel picture. Remember that to "Mark as Disturbing" and possibly delete, you need to be able to see the visible innards of the chopped-off head. But there's also his lifeless naked body in the picture. Do I delete the image for Nudity, Graphic Violence, or Dangerous Organizations? This is going to slow me down because I need to think.

Dangerous Organizations is also in the hierarchy, but I don't see any symbols from the cartel, and they usually like to include them. What's on the daily to-do list of a typical cartel member? Film an execution, make sure you torture him, and remember to include the gang symbols. Does the cartel have to deal with Quality Assurance on their executions? You filmed the execution, you did torture him, quite well, I might add, but you forgot to include the gang symbols. I think we need to put you on a Performance

Improvement Plan (PIP). Nudity was higher than Graphic Violence, so I think I deleted that one for Nudity, even though his genitalia were missing. It seems absurd that Nudity is of greater concern than Graphic Violence or Dangerous Organizations.

I hoped that job wouldn't be one of those reviewed by Quality Assurance. Audits were a pain. Out of maybe fifteen hundred decisions, they audited about fifty a week. If I dropped below 95 percent, I no longer met the Service Level Agreement, which happened at least once to pretty much everybody I knew. If you failed a couple of audits in a row, you got a Performance Improvement Plan. I could dispute the assessment, and often did, and most of the time I got my points back. But were they really checking me, or just seeing if I could defend my decisions? And what about those moderators who were too shy to stand up for themselves?

If I wrote a very well-crafted argument in my dispute, Quality Assurance was more likely to give me my points back. But then I was taking time away from my productivity. How is it that I was more concerned with my audit score than the human tragedy that flashed before me every day on the screen? But that was the system that had been set up, and like a rat in a maze, I was just trying to make my way through it. I think about some of the reviewers, go through a mental check-list of those I think like me, and those who seem to have a problem with me. The power Quality Assurance had over our lives seemed petty and tyrannical.

Eventually, management would remove our names and mask them from the audit process, something I had petitioned for. Before they became blind audits, Quality Assurance would be able to see the name of the person they were reviewing. But in a situation like this—where personal opinions could color actions—anonymity was important, which is why I petitioned for it.

There was so much gray area, and it was easy for Quality Assurance to mark a post as "ambiguous" and then get on to ruining another content moderator's day. They got paid a dollar an hour more than we did, the job wasn't as difficult, and it was harder to fire them than a content moderator. We were like rats fighting over who gets a marginally bigger piece of cheese.

I worked on complex objects for a long time, which gave me a break from needing to make three hundred decisions a day. These were groups and pages that were reviewed at Cognizant. If a third of the random posts were violating terms and conditions, then the whole group would come down. I recall feeling it was kind of a power trip to possess what we called "the ban hammer." We joked it was even more powerful than Thor's mighty hammer from the *Avenger* movies; "You are BANNED from Facebook as if you NEVER EXISTED!"

There were also live videos, normal videos, posts, and messages. Instagram, Facebook, we got it all. We were the Facebook security guards and prided ourselves on doing what the client asked.

They told us everything we needed to do.

They controlled the policy, the standards, the interpretation, even the interior decoration of the office. If I pushed back against a certain policy, I'd often be told to "use your head, not your heart" or "it's what the client wants."

Remember, they were the client. We worked for Facebook.

In May 2017, about ten months before I started working at Cognizant, *The Guardian* newspaper in Great Britain published a long article reviewing the rules under which Facebook was moderating content. It gives an insight into how Facebook was trying to deal with these issues:

> Facebook's secret rules and guidelines for deciding what its 2 billion users can post on sites are revealed for the first time in a Guardian investigation that will fuel the global debate about the role and ethics of the social media giant.
>
> The Guardian has seen more than 100 internal training manuals, spreadsheets and flowcharts that give unprecedented insight into the blueprints Facebook has used to moderate issues such as violence, hate speech, terrorism, pornography, racism and self-harm.[1]

I've heard it said that if you want to truly understand a subject, you should be able to put it all in a single-page memo, or give it to a person in what's called an "elevator pitch" in just a few minutes. The simple fact that Facebook had more than a hundred internal training manuals, not to mention countless spreadsheets and flowcharts, should make it clear to anybody that Facebook was not meeting its own "burden of clarity" for content moderation. The article continued, and one couldn't help but feel some sympathy for Facebook, which in many ways had become a victim of its own stupendous success:

> One document says Facebook reviews more than 6.5m [million] reports a week relating to potentially fake accounts, known as FNRP (fake, not real person).

Using thousands of slides and pictures, Facebook sets out guidelines that may worry critics who say that the service is now a publisher and must do more to remove hateful, hurtful and violent content.

Yet these blueprints may also alarm free speech advocates concerned about Facebook's de facto role as the world's largest censor. Both sides are likely to demand greater transparency.[2]

The article continued with some explanation of the conflicting rules as they struggled to define what could be allowed:

Remarks such as "Someone should shoot Trump" should be deleted, because as a head of state he is in a protected category. But it can be permissible to say: "To snap a bitch's neck, make sure to apply all your pressure to the middle of her throat", or "fuck off and die" because they are not regarded as credible threats.

Videos of violent deaths, while marked as disturbing, do not always have to be deleted or "actioned" unless there is a sadistic or celebratory element.

Photos of animal abuse can be shared, with only extremely upsetting imagery to be marked as "disturbing."[3]

It's interesting to note how Facebook would eventually change its policy regarding President Trump to allow images as extreme as showing his throat sliced under the rubric of political discussion. But at least when they started looking at these issues, it seemed Facebook understood what reasonable rules might look like, even though later they abandoned that standard. And while your average newspaper or television news show might talk about a death, or even show disturbing videos of violent events, there is usually a warning given or some filter to the information to prevent it from being especially graphic. The list of topics and the acceptable rules continued:

All "handmade" art showing nudity and sexual activity is allowed but digitally made art showing sexual activity is not.

Facebook will allow people to livestream attempts to self-harm because it "doesn't want to censor or punish people in distress."

Anyone with more than 100,000 followers on a social media platform is designated as a public figure, which denies them the full protections given to private individuals.[4]

Whenever an organization encounters problems for the first time, you have to give the decision makers some leeway. Human beings will make decisions they later realize were wrong. If I were in a decision-making role, I'm sure some of my decisions would be bad. These situations are difficult to balance. I was pleased that the article accurately depicted the challenge encountered by Facebook, although I may have disagreed with some of Facebook's conclusions:

> In one of the leaked documents, Facebook acknowledges "people use violent language to express frustration online" and feel "safe to do so" on the site.
>
> It says: "They feel the issue won't come back to them and they feel indifferent towards the person they are making the threat about because of the lack of empathy created by the communication via devices as opposed to face to face.
>
> "We should say that violent language is most often not credible until specificity of language gives us a reasonable ground to accept that there is no longer simply an expression of emotion but a transition to a plot or design. From this perspective, language such as 'I'm going to kill you' or 'Fuck off and die' is not credible and is a violent expression of dislike and frustration."[5]

Sometimes you think you're solving the problem, but then you realize you weren't even looking at the real problem. It's understandable that one might think the problem was social media leading to violence. But then you say to yourself, *These people aren't getting out of their seats to interact with people who may literally be feet or yards away from them.* Are they going to expend the energy necessary to get into a car, find the person's address, and then physically harm them? It seems like a lot of work. Maybe the real danger is that you're creating new, and less human, ways to communicate online. While I realize Facebook is unlikely to post guidance from the Mormon Church on profanity, they might want to create some similar document that incorporates those principles as a way to restore some civility on their platform.

* * *

In late 2018, that's why I asked to see what we could do to restore a post by Seth Gruber, a pro-life activist, that was deleted. I came across a different post that said his pro-life video had been deleted by Facebook. I investigated the problem and thought the complaint against the deletion was valid.[*]

---

[*] For a more thorough analysis of the issue with Seth Gruber, see Chapter 14.

I raised this as a possible public relations or PR "fire," which was one of our job responsibilities. If his post did violate Facebook policy, couldn't we make a "newsworthy exception"? Would the pro-life community, a significant segment of our population, feel that this was an abridgment of their First-Amendment rights? They have made the argument before that while their videos of abortion are disturbing, the same logic would have prevented videos of slavery being shown if the Internet had been around in the 1800s.

I was told that for Facebook to go into his account and restore it would violate Facebook's privacy policy. That's right.

Continuing the ban respected his privacy.

But going into his account to restore it would be a privacy violation. That's a little like the kid who kills his parents and then throws himself on the mercy of the court because he's an orphan. In this case, Facebook is like the kid who is using the same system that would try him for murder in order to gain reprieve from punishment. Facebook created their own privacy policy, then broke the rules, and are now saying they can't break the rules again to fix the mess they created.

The community standards at Facebook, or what we referred to as the "policy" (also known as the Implementation Standards [IS]), was a very complex legal document. Every question imaginable is spelled out in the most excruciating detail.

For example, we had a fairly common image that showed Melania Trump from the 1980s posing nude in bed snuggled up next to another woman, in a position many refer to as "spooning." Our "Nudity and Sexual Activity" section of the policy doesn't allow implied sexual intercourse. So, if someone is wearing clothing next to someone who isn't wearing clothing and it was crotch on crotch, even though we can't see visible penetration, it's counted as implied sexual intercourse and deleted. However, in this case they were both completely naked from the waist up, and there were no visible genitalia.

Technically, there couldn't be penetration, since they were both women. We raised this job up for guidance. The directive came back to delete this for implied sexual intercourse, which didn't make sense. It's decisions like these that are important because they fill in "gaps" in the policy until Facebook modifies their policy for cases like these.

The policy team at Cognizant told us how to act, but it was with input from Facebook. It was important for all of us to be aligned and to make decisions the same way for the sake of our site score. We needed to have Quality Assurance and the reps all on the same page.

The Cognizant policy team was made up of Cognizant employees who provided guidance, but oftentimes they would raise the question to Facebook for a final guidance decision.

Facebook employees generally visited our office about once a month. However, the Cognizant policy team would videoconference with them on a daily basis. The policy was constantly changing, and every two weeks we would have a policy update. It could be just a few tweaks, or it could be a major overhaul or reorganization. For example, the bullying policy used to be separate from the harassment policy. But they merged the two sections of the policy.

They also had a "hierarchy of actions" that they modified from time to time. For example, deleting terrorism was very high on the list of about thirty categories. Deleting child porn was also very high.

However, you could only delete something for one reason. If there were multiple violations, you would delete it for the highest violation. So, if there's a gruesome video of a car accident where you can see an arm fly off, but in that same video there's also nudity, the entire video would be deleted for nudity.

Again, how was this reasonable? Yes, I know that sometimes you enact policies that, when you go to implement, you realize don't make sense, but then don't you quickly adjust it? Isn't that supposed to be one of the benefits of technology, that you can quickly fix mistakes and policies that don't work?

Facebook changed the hierarchy around 2019 and made "hate speech" higher than "bullying." Prior to 2019, if I posted a comment attacking someone named Frank and called him a "piece of shit Christian and a whore," this would be deleted for bullying, because I'm calling him a whore. The attack on his Christianity wasn't taken into account for reporting purposes.

However, after they changed the hierarchy, hate speech was higher on the list. That same attack would be deleted for hate speech (the attack on being Christian) instead of bullying. The effect of this policy change resulted in more reports of hate speech on the platform.

The numbers would show that more hate speech was occurring. But really they just modified their rules for how it was reported. An increase in reports of hate speech would also give Facebook more motive and latitude to create a stricter hate speech policy.

* * *

Another fascinating change in their policy was with regard to attacks on "Public Figures": Prior to 2019, you could call a public figure a "whore," "bitch," or other female-gendered curse words (cunt, pussy, etc.) on their official page, and it was allowed. The new policy change, called "Purposeful Exposure," made it illegal to call public figures these names if they are directly mentioned (e.g. @alyssa_milano you're a bitch) or if it's a comment on their official Instagram or Facebook page.

To illustrate the absurdity of Facebook's rules, here's a good example. Alyssa Milano was allowed to say, "Men shouldn't be allowed to make rules about women's bodies," which clearly violates Facebook's hate speech policy for exclusion. Facebook gave a newsworthy exception for that phrase during Hollywood's campaign against Alabama's late-term abortion ban.

However, a Middle American nobody on Facebook can't call Alyssa a bad name on her Facebook page. Because that's too offensive for Facebook, and the policy literally mentions that the public figure might see that comment and be emotionally damaged by it.

Give me a break.

If you're in the public spotlight and can't handle a few verbal attacks, then quit. And besides, Alyssa Milano gets specific exemptions from Facebook's hate speech policy, something ordinary users of Facebook do not receive. Apparently, what she says is "newsworthy" and Facebook's rules for mere mortals don't apply to her, as will be shown in Chapter 14 of this book.

Going back to "Purposeful Exposure," this modification of the policy severely limited petty attacks on public figures. But it only disallowed attacks of claims of sexual activity, asking to engage in sexual activity, and female-gendered cursing. You can't call Alyssa Milano a "bitch" or a "cunt," but you can call any male public figure a "dick."

Male-gendered cursing wasn't mentioned at all with this policy change. I'm not in favor of any cursing, but if we're going to have rules, let's make sure they're the same for both sides.

Calling someone a "dick" is considered targeted cursing, so that would require a name/face match (NFM) from the person being attacked (private individual) in order to delete it. So, attacks on women are *not* allowed, but attacks on men *are* allowed. Ten points for feminism.

This all goes to show how nuanced the policy is, and how small changes can make a big impact. It also illustrates how Facebook has full editorial power over their policy and can give exceptions willy-nilly in order to promote certain movements.

The key takeaway from the Purposeful Exposure change to the Bullying policy is how it increases how often we see these pages. Facebook was blind without us. We were their eyes and ears. They didn't know the trends or common topics without content moderators. So, by tightening up attacks on public figures, they would have greater visibility into the kinds of conversations happening on the platform.

This ties into the "Civic Harassment" queue, which was a new queue designed for the 2020 election. This was yet another policy change that was designed to ensure a fair and democratic process for the election.

Considering what happened, you'll forgive my skepticism.

This is from a screenshot of their "Civic Harassment Training PowerPoint" that I was shown:

> Why were the changes made to the existing GRT [guided review tree] for B & H [Bullying & Harassment]?
>
> Bullying and harassment has been identified as a priority issue around the US 2020 election. We acknowledge that anyone can share an opinion about the US 2020 election, but not all voices carry equally far nor are equally susceptible to attacks. We want to protect not only influential figures who are vulnerable to harassment through their status, but also ordinary folks that make themselves vulnerable by interacting with content generated by these figures.
>
> We've also identified that we will have quite an overlap with hate speech content while working on this effort—and we would like to measure that overlap by having a different guided review tree than the 3-step labeling tree.
>
> This is also an important focus area for Civic Integrity and they will be using the results from this work for their overall elections work.[6]

Before 2016, Facebook wasn't so frenetic and frantic about election integrity. I'm sure they had policies in place, but before 2016, the large majority of their content moderators worked overseas.

What we were told at work was that Facebook acknowledges there was Russian interference in the 2016 election, along with voter fraud. That was the principal reason behind their hiring thousands of US-based content moderators. Companies like Cognizant and Genpact cashed in on this opportunity. The agreement between Cognizant and Facebook was a three-year, two-hundred-million-dollar contract.

So why the urgency?

Why did Facebook urgently need visibility into nonviolating trends about the Democratic debates? I believe they wanted this information so they could, at a later date, take action.

The 2018 US midterm elections were great practice for Facebook.

By that time, they had a sizable content moderation team at my Phoenix location with roughly a thousand employees. The training for the 2018 midterms was very extensive. During this period, I was on the Spanish language side, but I still had access to the election training decks for the United States. And on occasion I would moderate posts in English if the queue was backed up. For the last seven months of my time at Cognizant, I was solely working the North American English queue, from approximately July 2019 until February 2020.

Some of the election training included deleting posts with the wrong date for the election and marking with "VI" (Voter Information) content related to the election. For example, if a post was encouraging someone to vote, we would type "VI" in the notes so that Facebook's election team could review it.

Any mention of "urging people to vote, providing instructions for voting, or giving out general information about the voting process" would also be marked with VI and would be flagged for review by Facebook employees.

In the winter of 2019–2020, we saw an uptick in posts about civil war, impeachment, and boogaloo. As I was trained to do, I raised these trends to my supervisor and sent an email to COPHXTRENDS@cognizant.com.

The Phoenix policy manager, Shawn Browder, was pleased that I was raising those trends. Here is a transcript of our conversation:

2:30 **Ryan:** Is the boogaloo serious, are they joking?

2:41 **Shawn:** Some people are joking, some people are serious. The danger of leaving it on the platform is, where's that line.

3:11 **Shawn:** Definitely the civil war stuff.

Paul[†]: You can see people hyping each other up.

3:19 **Shawn:** That's exactly the type of stuff that Facebook wants to see, like had they seen that type of stuff, like leading to the 2016 election, like they would have definitely like put some things in place, to prevent it.[7]

---

† A pseudonym.

This further reaffirms the fact that Facebook was surprised by the 2016 election results. Their reaction to the 2016 election was to initiate mass hiring of US content moderators, create election training decks, and modify the queue for the 2020 election. So, we know the election was a high priority for Facebook.

I agree that preventing election fraud is important. And on its face these are innocent actions and we should support such efforts. However, what I witnessed is that whether intentionally or inadvertently, Facebook's political slant seeped into its policies. Whether this was blatantly intended to prevent Trump from winning is a question we all need to ask.

* * *

I remember the day my coworker struggled with some of the content he had to see. Jim‡ confided in me that something he saw at work really bothered him.

He had reviewed some disturbing child pornography and hadn't slept for the last couple of nights. Jim had a young daughter, and seeing such graphic imagery and explicit sexual content was haunting him. He couldn't get the image out of his head.

I discreetly reached out to the counselor on site, Alan,§ so that Alan could talk to Jim.

I did this on countless occasions. It was something many of us did for one another. Sometimes you're not the best judge of what you need. I include myself in that category. If I noticed someone was having a bad day or seemed out of sorts, I asked if they were okay and reminded them about the counselors on site. This wasn't a defined role, but I felt like it was my duty as a member of our work community.

When you experience traumatic events together such as these, it forms a special bond. I felt like there was a special camaraderie among our content moderators. Putting aside the political partisanship of Facebook for a moment, we did help keep the Internet a little cleaner than it was before.

But there's a price to be paid to keep the Internet safe. We took the brunt of that burden, and it affected our personal lives, our marriages, and our relationships.

Graphic imagery affects everyone a little differently, regardless of where you work.

---

‡ A pseudonym.
§ A pseudonym.

In 2011, I worked at a funeral home for a few months. I never got used to seeing a dead body. The only thing I got acclimated to was grabbing the legs of the deceased, but nothing beyond that. I remember I was in the freezer with my coworker when he had to remove the pacemaker from a loved one. He essentially stabbed the chest and pulled out the pacemaker. Things like this affect one's psyche.

That image haunted me for days.

Again, in 2014, I witnessed a dead body while working security at a grocery store in Phoenix, Arizona. I was doing a patrol of the parking lot and thought someone was sleeping in their car. When I went back before my shift ended, I noticed he hadn't moved and there was discoloration on his hand. I called 911 and was traumatized by this event. I stood rooted to the ground for about an hour while the police questioned me.

I think those experiences toughened me up and made it easier to deal with the images that came across my screen.

Going into my role as a content moderator, my background was probably a bit different than most. I was thirty years old, which was old compared to most of the workers. The average age was probably twenty-three.

For those who have preexisting emotional issues, this was not the job for you. In my case, I feel like I was somewhat desensitized because of my previous jobs. But the first couple of months on the floor were shocking.

It would be nice if Facebook's AI told us what we were going to see before it popped up on our screen. But you literally clicked "next," and anything could appear on the screen: child porn, a cartel beheading, animal "crushing," bestiality, terrorism, rape, incest. You found yourself wondering, *How many monsters are really in our society?*

My last day working at Cognizant was February 28, 2020. I'm blessed that I don't have recurring nightmares or images that pop into my head. While working there, I did my best to not think about work when at home.

When I came home from work, I'd play NBA 2k20 on the Xbox to decompress. Ryan Hartwig for the three-pointer at the buzzer! Score! I also like to read science fiction and did that before bed. I guess it's similar to how surgeons are instructed to think of their work. Their patients aren't people. They're pieces of meat, which need to be carved up and made better.

To this day I'm not sure if that's the best way to maintain your humanity.

* * *

In our training group, we were constantly exposed to pornographic images.

It was difficult at first to not stare at the image and focus instead on which elements of the policy the image violates. But you have to study the image to see what it violates. Believe it or not, after a couple months, you get desensitized to it.

Obviously, visible penetration was an infraction. But there were more nuanced examples of near nudity, and the policy was very specific. Underwear advertisements were given a pass. Those damned male models get away with everything! Not that I'm jealous or anything.

For a long time, nudity in a birth-giving context was against the rules. But Facebook eventually adapted their policy to allow for it. Also, nipples in a breast-feeding context are allowed. We would mark imagery that didn't violate the rules, but that had sexual elements, with the "CE" label. This stands for Continuous Enforcement and was how we helped Facebook train its AI.

This was to allow users to control their content. For example, if someone didn't want to have any scantily clad women on their timeline, they could control what they want to see. Cleavage was also something we had to look for and identify. In a video review, we had to look in the thumbnails and recognize any cleavage. I think this affected content moderators in how they viewed women and caused objectification of women.

We also had to mark with CE anytime we saw male "bulge" in thumbnails. Of course, we had a mixed group of both male and female content moderators, but I think this also objectified men.

* * *

One of my coworkers was an example of someone who probably shouldn't have started working there in the first place. Going into the job he had anxiety issues, and the stuff he saw at work only exacerbated this. It's unfortunate that this can happen, and I wish there were more safeguards in place to prevent at-risk individuals from engaging in this type of work.

In addition to your breaks, you had an extra fifty minutes of wellness time a week that you could use anytime. The Wellness Team also planned fun activities weekly, where we could wear different themed t-shirts or costumes. There were team activities and competitions that allowed us to go outside and do yoga or play wiffle ball. It was a fairly relaxed environment, and the dress code was very casual. I would wear jeans and a t-shirt to work, and I think all of my coworkers enjoyed being able to wear comfortable clothing.

Working for Cognizant had many benefits. Since Cognizant is a larger company (actually, a Fortune 200 company), the medical benefits available were much more substantial than if it were a smaller company. The health insurance was quite good, and there were many other perks that you usually don't see, including up to three free in-person counseling visits outside of work, per issue, per year.

Human Resources also had a difficult role. Many employees were younger and constantly complained about trivial work conditions outside of Cognizant's control. Yes, these issues are important, but employees would hijack our group meetings (town halls) to voice individual concerns. For example, the parking garage didn't have cameras, and cars were being broken into. This is the responsibility of the property management company. This issue was raised by Cognizant, but the property management company did nothing.

Since we were a 24/7 facility, the number of issues at work increased. We had vandalism from employees in our wellness rooms. Cognizant purchased massage chairs and transformed a few small offices into relaxation rooms. One of the chairs was slashed with a knife. We also had graffiti in the bathroom. At one point early on, there were rumors of sex in the stairwell. Is it surprising for that to happen when you have a bunch of twenty-somethings watching porn at work all day, as part of their job?

Once again, it's common for those who've never spent much time outside the United States to bitch and complain about work conditions. None of them have lived in a third-world country, like I did in Chimalhuacan, outside of Mexico City, in the year 2007. I had to heat my shower water by sticking an exposed live wire into a bucket of water. The streets weren't paved, and I had to wet down the concrete outside my front porch to keep the homeless from sleeping there. I met people who were so poor they had to inhale paint thinner to keep from starving. Clean water was a luxury.

Yes, there is always more a company can do to help its employees. But by my measure (and I've worked at more than twenty jobs), Cognizant did everything in its power to help their employees and advocate for them.

* * *

Working as a content moderator was a unique job. The types of discussions we had ranged from the bizarre to the abnormal to the extremely awkward. As a community, we did support one another and form a culture that in many ways was positive.

When I was on the Spanish side (ESLA), I felt like the culture was better. As I mentioned, I was one of two "gringos" and got along well with everyone. I was on the Spanish side from when I started in March 2018 until July 2019, and then I transitioned to the North America side due to severe attrition. A lot of people were quitting and going to work in customer service for DoorDash.

So, when I transferred to the North America side in July 2019, I did notice some cultural differences. Apart from having to learn some of the different nuances as far as "actioning" content⁵ went, I noticed that the morale was a little lower on the North America side.

I transferred over to North America with a few people from the Spanish side, including a guy named Jose Moreno. He is seen in the June 25, 2020, Project Veritas video release. He's a military veteran and pretty blunt, but I came to appreciate his sense of humor. He brought a lot of positive culture to the North America side. Jose and Tyrell Lease were both military veterans, and in the last couple of months of work, I sat near both of them.

Tyrell was also featured in the video from Project Veritas. Tyrell and Jose are both conservatives and were able to talk openly about conservative issues at work. I'm pretty certain there were other conservatives at work who weren't as brazen and bold in expressing themselves as Tyrell and Jose, but I didn't know them.

The last couple of months of work, in January and February 2020, the environment was very relaxed. We all knew we were getting laid off at the end of February due to Cognizant's choosing to end the contract half a year early. We suspected the contract cancellation may have been due to the negative publicity surrounding Cognizant after Casey Newton's 2019 article was released, which portrayed Cognizant in a damaging way (we will cover this in more detail later).[8] I imagine it was like serving on a Navy ship that you knew was getting ready to be decommissioned. Facebook still put jobs in the queue, but we'd spend like five minutes on each job. Under normal conditions we were supposed to "action" each job in thirty seconds. So, toward the end, each person was "actioning" maybe fifty jobs a day.

I was always kind of a stickler when it came to that, and I would still try to action two to three hundred jobs a day.

When I got to the North America side in July 2019, I also developed a similar reputation as a straight-laced, hard-working employee. Some of my

---

⁵ As content moderators, we would refer to the review of content as "actioning" content. Because we would "action" one job at a time, every time something appeared on our screen. That's just the lingo we used internally to describe our daily job duties.

colleagues joked I was the "undercover boss," based on the popular CBS show *Undercover Boss.* I'd gone from being the "gringo" on the Spanish side to "the boss" on the America side.

Apparently, I have a certain look about me, a straight-shouldered, plain-spoken, no-nonsense CEO type. I'd joke to my coworkers that yes, I was the undercover boss, so they'd better behave. If only they knew what I was really doing.

Because by that time, I had taken up the Project Veritas challenge to "Be Brave! Do Something!" and was filming undercover, documenting the liberal bias of Cognizant and their client Facebook.

CHAPTER FOUR

A Tale of Two Whistleblowers

Although the term *undercover* is usually used to describe methods and actions performed by law enforcement, I was for all intents and purposes "undercover." I remember telling one of Project Veritas's undercover journalists, "This is like some spy shit, huh?"

I did NOT get the job at Facebook with the intention of sending recordings to Project Veritas. I'd heard of them, but mainly knew about the undercover journalism from Live Action, a pro-life organization that is often confused with Project Veritas.

In February 2019, I had been at Cognizant/Facebook for almost a year, and that month, when Casey Newton's article came out, titled "The Secret Lives of Facebook Moderators in America,"[1] Casey came to our Phoenix office, and I actually saw him in the hallway. The focus of that article was the mental health of content moderators. However, I think subconsciously this made me more aware of other issues at work.

In the first year, I had been mentally cataloguing a few aspects of the policy that seemed weird to me. After Casey Newton's article came out, the subsequent months are when I finally put a few of those discrepancies on paper.

So, in May 2019 I wrote a list of nineteen examples and mailed it to some members of Congress and a few US senators, including Congressman Paul Gosar, and also Senator Marsha Blackburn. I believe I sent a copy to then-Senator Martha McSally. The letter is dated May 14, 2019, and has a list of nineteen examples of bias from Facebook, many of which I've already mentioned, such as my trainer in March 2018 speaking openly of Barack

Obama as her "Patronus Charm," as well as being allowed to call straight white males "filth" if they didn't support the LGBTQ agenda. However, I provided additional examples of what I believed were bias:

> In summer of 2018 there was that viral video of a teenager in a fast food restaurant getting his Trump hat knocked off. We were specifically told to delete all instances of that video instead of making a #newsworthy exception.
>
>   In April 2018, Facebook made a #newsworthy exception to allow a video of a Palestinian child being killed by an Israeli sniper, in spite of their Graphic Violence policy.[2]

Have you ever found yourself wondering why certain stories catch fire and get attention, while others don't? The fact is, it's just like a fire, and you need to give it oxygen. In this instance, the oxygen to start the fire is allowing the video to be seen. If the video simply vanishes from social media, it's as if it doesn't exist. No fire starts.

How could we talk about violence against Trump supporters in the summer of 2018 if the video clearly showing an example of such behavior just vanished from social media? However, in the spring of 2018, everybody was talking about the video of that Palestinian child being shot by an Israeli sniper.

One of my concerns was that Facebook was using content moderators at Cognizant who were supposed to be monitoring content from the United States to affect the politics of foreign countries. As I wrote in my letter, "There was a specific request from Facebook in SRT (Single Review Tool) for us to flag right-wing extremist groups outside the U.S."[3] Why were we in the United States so concerned about groups outside the United States?

Another strategy used by Facebook was to specifically target certain individuals in an attempt to lower their profile, even though they hadn't been convicted of a single crime. One such individual was Gavin McInnes, founder of the Proud Boys:

> Gavin McKinnis [sic] (1174 objects removed by Facebook) and [sic] added numerous people throughout the world to their hate org list, mainly anti-immigrant individuals in Europe. The last thing that pushed me over the edge (this past weekend) was them listing "white nationalism" as a designated hate organization, and they specifically said in their video of the policy rollout, that it only applies to "white nationalism and separatism" and not any other race. So, white nationalism has a special designation but black

separatism is fine. Also if you search "black power" and then "white power" in SRT you'll find that only "white power" is violating.[4]

Do you see how easy it is to lower somebody's media profile if you simply say they're part of a "hate" organization? In my opinion, Facebook made the clear-eyed decision that Gavin McInnes needed to be removed from the public discussion. This is dramatically affecting the public debate. Personally, I don't believe in Black power, Asian power, or White power, but the power of the individual. The rules need to be the same for every group. Otherwise, you're favoring one group over another. People should be grouped together by common principles, not the color of their skin or ethnic background.

Did Facebook understand how perilous a course they were undertaking? I believe they did, because it seems they were also taking steps to ensure the ideological purity of their content moderators. As I wrote in example sixteen of my letter:

> Also, I don't know if I mentioned this before, but last summer (2018) they required all content moderators to link their personal Facebook accounts to the project. They said it was so that we didn't action content from our friends, but it was kind of suspicious. We just logged in once at work and they said it only saw who our friends were in order to not action their content.[5]

Did I believe that Facebook just wanted to see who our friends were? I do not. I think they were probably keeping that information to check up on us and see if we deviated from the liberal orthodoxy. If I'm right, I expect that information will eventually make its way into the public domain, if Facebook allows it.

* * *

I'm somewhat impatient, and so later in May or early June 2019, after I had mailed the letter with nineteen examples of bias to a few members of Congress and the US Senate, I reached out to a few news organizations to see if they had any interest. One journalist suggested I should reach out to Project Veritas.

I used an encrypted messaging app and sent a message to their tip line. I'm not sure of the exact time line. But not too long after that I got a phone call from someone with Project Veritas. They'd been VERY interested in Facebook. I'd learn later that they also had another whistleblower.

In the conversation with Project Veritas, I offered to film with a hidden camera. I'm not sure what made me so brave in this initial conversation. For better or worse, I think I have a moral compass that says if I believe something is wrong, I have to stand up and take action. They subsequently sent me a camera to use, and I began filming, sporadically at first, then more regularly. I didn't know much about journalism. My only background in reporting things was as a security guard, but I did have a long history of writing up "incident reports" and making sure events were accurately reported. I decided to treat the situation as one big incident report. As a content moderator, I'd essentially been hired as a security guard for Facebook. But now I felt I was defending free speech, a right more valuable than any company or industry.

After a day of filming, I'd go home and transcribe the audio and send it to Project Veritas. I didn't initially tell my wife that I was filming. Later she told me she had found the camera at home but didn't mention it to me.

But around November of 2019, I fully explained to her what I was doing. We had a difficult conversation about it, because the job had good benefits, even though I was only making fifteen dollars an hour. And to be completely honest, I did enjoy the mental challenge of the job and the culture, once you got past the graphic content. Cognizant is a Fortune 200 company, so the health benefits were top notch, in addition to having a 401k.

If I hadn't filmed and gone public, I'm certain I could have stayed on at Cognizant and transferred to another project fairly easily.

However, I don't regret filming with a hidden camera, even though I sacrificed the possibility of a steady career. I'm just now realizing the scale and importance of what I uncovered. It didn't seem like much at the time, but Big Tech censorship has become an even more important theme in the last few months while I was writing this book. I expect that to continue in the years to come as we either rediscover our freedom as Americans or go down the dark road of dictatorship.

* * *

I recently began watching a Netflix show called *The Social Dilemma*. In it, many former executives of tech companies talk about the harmful effects of social media and the business model. They describe the business model as one focused on manipulating one's behavior to spend more time on social media.

In this regard, Facebook is not merely a platform, but rather an agent of change. It can influence behavior. If their whole business model is based on causing people to change their behavior, why are we allowing them to be involved in elections?

I appreciate the honesty of the executives featured in this show. However, I'm worried that they'll come to the conclusion that Facebook needs to do more to fight hate speech. That they need to take more action against conservatives, while Antifa and other violent leftist groups are given free rein on the platform.

Many of these individuals profiled in *The Social Dilemma* are essentially technology ethicists. They're concerned with where Big Tech is headed. But that's the danger of technology. You can't control how it's used. It's unpredictable by its very nature.

Those who argue social media should only be used for good would end up arguing for more limitations on free speech. Additionally, those who argue for more limitations are arguing that Facebook is not a platform, but a publisher.

So yes, Facebook manipulates human behavior to an extent we've never seen possible. And they influence roughly a third of the world population. It's one thing to create a platform.

But it's something else altogether to actively manipulate, poke, prod, and encourage users toward certain actions and behaviors. Based on Facebook's abysmal history of data privacy and manipulation, I find it very troubling to see that they're involved in tracking and monitoring elections worldwide.

In this documentary, the question posed is "What does Facebook do with these data?" They are using their wealth of data to manipulate and change societies. They even want to know what's trending, even if it doesn't violate their standards.

The real dilemma now is how we act against Facebook. They wield absolutely too much power, and at this point Congress and other countries are powerless against their monopoly.

The science fiction novel *Caves of Steel* by Isaac Asimov portrays a similar type of juxtaposition between technology and society. Yes, technology can be used for good. But at what cost to society? What are the impacts of this new reality interwoven with technology? Oftentimes, we can't tell what is real and what is manipulation from Big Tech. In *Caves of Steel*, we see the same problem—a robot AI that looks and acts like a normal human.

How much of Facebook is real and how much is packaged information designed to nudge us toward a preapproved opinion?

Let me tell you two stories of people in the news, what I saw behind the scenes at Cognizant/Facebook, and what you, the news-consuming public, saw.

* * *

Since the Project Veritas video release on June 25, 2020, I've done over a hundred interviews, and a lot of people have asked about the jobs I reviewed. Many are curious to know how the job got to our screens: is it the AI, are there separate queues?

Are the jobs we get random, or does Facebook choose what we review?

The short answer is it was a mix of all these approaches.

At Cognizant, we began with just was-live or prerecorded videos, then gradually expanded as Facebook gave us other types of work. Another queue is CM or content moderation, which included comments, posts, and direct messages on either Instagram or Facebook. Many of these posts are user-reported. The only reason they arrive in our queue is that someone reported them.

There were also a lot of posts that are reported by bots, and we reviewed those, as well.

As far as queues go, there was also a queue just for hate speech edge cases and IGPR, which is Instagram Profile Review. A separate queue was Complex Objects, where groups and pages were reviewed. Live videos were also another line of work.

We'd get our jobs in batches. Facebook would dump ten thousand jobs on us and pay us for taking action on them. While I was on the Spanish side, I was gradually trained in many lines of work. I'd get an assortment of posts, comments, and videos, as well as live videos.

During the entire time in Spanish, I had access to the North America posts on SRT, and we occasionally actioned jobs in English, as well.

When I transferred to the North America queue, I was taken off the live video queue but was still in the CM queue and reviewed posts and was-live videos.

While on the North America side from July 2019 to February 2020, one case in particular showcased Facebook's ability to prioritize certain jobs to place them in our queue for review. These two examples deal with the person known to history as the Ukrainian whistleblower.

* * *

The Ukrainian Whistleblower is the individual who heard a secondhand account of a telephone conservation between President Donald Trump and Ukrainian President Volodymyr Zelensky. The National Security Agency changed the rules to allow for secondhand accounts of whistleblowers shortly before the whistleblower submitted his report.

This report eventually led to an impeachment inquiry and impeachment hearings against Donald Trump between September and December 2019.

The fact that Republicans couldn't say the name of the whistleblower, even during the Senate impeachment trial, is telling. This was a huge Republican talking point, and it would help their argument against the impeachment if they could discuss and mention the individual responsible for the launching of the impeachment inquiry.

Facebook also cooperated with Democrats and disallowed even the mere mention of this Ukrainian Whistleblower, whose real name is Eric Ciaramella. This is strange because it doesn't fit any of Facebook's policies.

Facebook disallowed a political talking point with absolutely zero policy rationale.

With regard to the phone call that Ciaramella blew the whistle on, President Trump called it a "perfect" conversation, but the news really didn't cover it. Here is a transcript of that conversation as released by the White House, but, of course, they didn't talk much about the actual substance:

> **President Trump:** Congratulations on a great victory. We all watched from the United States and you did a terrific job. The way you came from behind, somebody who wasn't given much of a chance, and you ended up winning easily. It's a fantastic achievement. Congratulations.

> **President Zelensky:** You are absolutely right Mr. President. We did win big and we worked hard for this. We worked a lot but I would like to confess that I had an opportunity to learn from you. We used quite a few of your skills and knowledge and were able to use it as an example for our elections. And yes, it is true these were unique elections. We were in a unique situation in that we were able to achieve a unique success. I'm able to tell you the following; the first time you called me to congratulate me when I won my presidential election, and the second time you are now calling me when my party won the parliamentary election. I think I should run more often so you can call me more often and we can talk over the phone more often.

President Trump: [laughter] That's a very good idea. I think your country is very happy about that.

President Zelensky: Well yes, to tell you the truth, we are trying to work hard because we wanted to drain the swamp here in our country. We brought in many, many new people. Not the old politicians, not the typical politicians, because we want to have a new format and a new type of government. You are a great teacher for us and in that.

President Trump: Well it is very nice of you to say that. I will say that we do a lot for Ukraine. We spend a lot of effort and a lot of time. Much more than the European countries are doing and they should be helping you more than they are. Germany does almost nothing for you. All they do is talk and I think it's something that you should really ask them about. When I was speaking to Angela Merkel she talks Ukraine, but she doesn't do anything. A lot of the European countries are the same way so I think it's something you want to look at. But the United States has been very, very good to Ukraine. I wouldn't say that it's reciprocal necessarily because things are happening that are not good. But the United States has been very, very good to Ukraine.

President Zelensky: Yes, you are absolutely right. Not only 100%, but actually 100%. And I can tell you the following; I did talk to Angela Merkel and I did meet with her. I also met and talked with Macron and I told them that they are not doing quite as much as they need to be doing on the issues with the sanctions. They are not enforcing the sanctions. They are not working as much as they should work for Ukraine. It turns out that even though, logically, the European Union should be our biggest partner, but technically the United States is a much bigger partner than the European Union and I'm very grateful to you for that because the United States is doing quite a lot for Ukraine. Much more than the European union, especially when we are talking about sanctions against the Russian Federation.

President Trump: I would like you to do us a favor, though, because our country has been through a lot and Ukraine knows a lot about it. I would like you to find out what happened with this whole situation with Ukraine, they say Crowdstrike. I guess you have one of your wealthy people, the server, they say Ukraine has it. There are a lot of things that went on, the whole situation. I think you're surrounding yourself with some of the same people. I would like to have the Attorney General call you or your people, and I would like to get to the bottom of it. As you saw yesterday, that whole nonsense ended with a very poor performance by a man named Robert Mueller, an incompetent performance, but they say a lot of it started with Ukraine. Whatever you can do, it's very important that you do it if that's possible.

President Zelensky: Yes, it is very important for me and everything that you just mentioned earlier. For me, as a president, it is very important and we are open for any future cooperation. We are ready to open a new page on cooperation in relations between the United States and Ukraine. For that purpose, I just recalled our ambassador from United States and he will be replaced by a very competent and very experienced ambassador who will work hard on making sure that our two nations are getting closer. I would also like and hope to see him having your trust and your confidence and have personal relations with you so we can cooperate even more so.

I will personally tell you that one of my assistants spoke with Mr. Giuliani just recently and we are hoping very much that Mr. Giuliani will be able to travel to Ukraine and we will meet once he comes to Ukraine. I just wanted to assure you once again that you have nobody but friends around us. I will make sure that I surround myself with the best and most experienced people. I also wanted to tell you that we are friends. We are great friends and you, Mr. President, have friends in our country so we can continue our strategic partnership. I also plan to surround myself with great people and in addition to that investigation, I guarantee as president of Ukraine that all investigations will be done openly and candidly.

President Trump: Good, because I heard you had a prosecutor who was very good and he was shut down and that's really unfair. A lot of people are talking about that, the way they shut your very good prosecutor down and you had some very bad people involved. Mr. Giuliani is a highly respected man. He was the mayor of New York City, a great mayor, and I would like him to call you. I will ask him to call you along with the Attorney General. Rudy very much knows what's happening and he is a very capable guy. If you could speak to him that would be great. The former ambassador from the United States, the woman, was bad news and the people she was dealing with in the Ukraine were bad news, so I just want to let you know that. The other thing, there's a lot of talk about Biden's son, that Biden stopped the prosecution and a lot of people want to find out about that so whatever you can do with the Attorney General would be great. Biden went around bragging that he stopped the prosecution, so if you could look into it. It sounds horrible to me.

President Zelensky: I wanted to tell you about that prosecutor. First of all, I understand and I'm knowledgeable about the situation. Since we have won the absolute majority in our Parliament, the next prosecutor will be 100% my person, my candidate, who will be approved, by the Parliament and will start as a new prosecutor in September. He or she will look into the situation, specifically to the company that you mentioned in this issue. The issue of the investigation of the case is actually the issue of making sure to restore the honesty, so we will take care of that and work on the investigation of the case.

On top of that, I would kindly ask if you have any additional information that you can provide to us, it would be very helpful for the investigation to make sure we administer justice in our country with regard to the Ambassador to the United States from Ukraine. As far as I recall, her name was Ivanovich. It was great that you were the first one who told me she was a bad ambassador because I agree with you 100%. Her attitude towards me was far from the best, as she admired the previous President, and she was on his side. She would not accept me as a new President well enough.

President Trump: Well, she's going to have to go through some things. I will have Mr. Giuliani give you a call and I am also going to have Attorney General Barr call and we will get to the bottom of it. I'm sure you will figure it out. I heard the prosecutor was treated very badly and he was a very fair prosecutor so good luck with everything. Your economy is going to get better and better I predict. You have a lot of assets. It's a great country. I have many Ukrainian friends. They're incredible people.

President Zelensky: I would like to tell you that I also have a quite a few Ukrainian friends that live in the United States. Actually, last time I traveled to the United States, I stayed in New York near Central Park and I stayed at the Trump Tower. I will talk to them and I hope to see them again in the future. I also wanted to thank you for your invitation to visit the United States, specifically, Washington DC. On the other hand, I also want to assure you that we will be very serious about the case and will work on the investigation. As to the economy, there is much potential for our two countries and one of the issues that is very important for Ukraine is energy independence. I believe we can be very successful and cooperating on energy independence with United States. We are already working on cooperation. We are buying American oil, but I am very hopeful for a future meeting. We will have more time and more opportunities to discuss these opportunities and get to know each other better. I would like to thank you very much for your support.

President Trump: Good. Well, thank you very much and I appreciate that. I will tell Rudy and Attorney General Barr to call. Thank you. Whenever you would like to come to the White House, feel free to call. Give us a date and we'll work that out. I look forward to seeing you.

President Zelensky: Thank you very much. I would be very happy to come and would be happy to meet with you personally and get to know you better. I am looking forward to our meeting and I would also like to invite you to visit Ukraine and come to the city of Kiev, which is a beautiful city. On the other hand, I believe that on September 1 we will be in Poland and we can meet in Poland hopefully. After that, it might be a very good idea for you to travel to Ukraine. We can either take my plane and go to Ukraine, or we can take your plane, which is probably much better than mine.

President Trump: Okay, we can work that out. I look forward to seeing you in Washington and maybe in Poland because I think we are going to be there at that time.

President Zelensky: Thank you very much, Mr. President.

President Trump: Congratulations on a fantastic job you've done. The whole world was watching. I'm not sure it was so much of an upset, but congratulations.

President Zelensky: Thank you, Mr. President. Bye-bye.[6]

By reading that conversation of twenty-seven minutes, you are now in possession of the same information as the so-called Ukrainian Whistleblower.

Did you find an impeachable offense in that conversation?

I didn't.

I heard two world leaders talking to each other with respect and trying to make sure that their respective countries were free of corruption.

* * *

My coworker Jaden* on November 6, 2019, had been getting multiple jobs relating to claims that the Ukrainian Whistleblower was either Eric Ciaramella or Alexander Soros, so he raised it up the chain. That is how we initially got the guidance from the Phoenix Policy Team with Cognizant. Specifically, Shawn Browder was shown the job outing Eric Ciaramella and decided it was to be a delete for privacy.

In the moment of Jaden getting the guidance, I filmed Jaden telling me about why it met for privacy violation. I also filmed Shawn explaining why he thinks it met the letter of the policy for the privacy violation. The privacy policy attempts to protect undercover law enforcement that is being outed publicly.

This interim guidance to delete under the Privacy policy was active for five hours, and then Facebook told us to delete it under the Coordinating Harm policy, effective at 7:52 p.m. on November 6, 2019. Myself and Jaden were the first ones to raise this issue, and before we raised it to Facebook, Eric Ciaramella was not being deleted.

Two days later, on November 8, 2019, a coworker named Alfredo† was on floor support helping us with jobs. While he was helping me specifically with a job exposing Eric Ciaramella as the whistleblower, I asked him if he thought Facebook was protecting the whistleblower. This is what I recorded and transcribed:

> **Ryan:** Hey what's up Alfredo? So, this is kind of like what we've been dealing with a lot. I'm getting a lot of this content. I'm collecting job IDs. So, I've got like, it looks like it says "the whistleblower has lots of connections in high places." We have a picture of him, right, so just the picture by itself would not be violating, right? Cuz, let me see, so we have "image of Eric Ciaramella accompanied with a whistleblower claim," so we have the image, and then it says whistleblower, right, so that's gonna be a delete for Coordinating harm > other
>
> So it's funny cause yesterday, um, Jaden is the one who first brought it up to Shawn's attention. So yesterday the post was like, oh it's privacy because they're undercover law enforcement officer. But how is he, how is he a whistleblower, how is he law enforcement if he's just a um . . .

Alfredo: Cause I believe he was part of the CIA, so I think they're just throwing CIA under the same thing as undercover, along those same lines, you know, cause like, if people are in the CIA, you don't really know.

Ryan Hartwig: So, he's kind of like law enforcement?

Alfredo: Yeah. This one [points at job on screen] I would say you have it [delete], you know clear cut, you have an image of him, accompanied with whistleblower claim, you just got it for that.

Ryan Hartwig: So, are they trying to like protect him then I guess? Is Facebook trying to protect him?

Alfredo: I guess, in a way, the whole theory man, conspiracy. [laughs]

Ryan: It's all political man.

Alfredo: It really is. That's really how, you know, Facebook deals with things. Did you see the whole Twitter's like not allowing like paid ads for like political? It's pretty cool.

Ryan: Oh yeah.

Alfredo: I like it. Like you can't be paid to have your stuff on there. Facebook ain't doing that though.

Ryan: Yeah I think they should just like delete everything. Not be involved in the business at all. Because I feel like they take sides too often.

Alfredo: For sure, and then like they have all these like algorithms and everything. This is the people [unintelligible] type of stuff, like they have data so they, I don't even know what they do with all this data.

Alfredo: Well, we know. We know we're helping them right now get the next president.

Ryan: Yeah, I know, right?

Alfredo: Which is weird.

Alfredo: [going back to the job on the screen] Yeah, naw with this one I would just go with the Coordinating [coordinating harm delete other] on it. There's your confusion on the whole project. Just go for that. That's what the client wants to do.

Ryan: Sounds good, Thanks Alfredo.[7]

Let's think about how this issue is playing out differently from most every other legal dispute in history. The cornerstone of our judicial system is that you get to face your accuser and know their name.

The reason is simple.

How are you supposed to prepare your response if you don't know who's making the claim against you? Is it your best friend or greatest enemy? Somebody you worked with, or somebody you've never even met? These things matter. We don't have "secret evidence" in the American justice system. That's more like a Kafkaesque, totalitarian nightmare.

But Eric Ciaramella was allowed to accuse the president of the United States of treason, and we, the American public, weren't allowed to say his name or speculate about his motives. It's often said that the president is not supposed to be above the law, but this clearly seems like President Trump was below the law that would apply to every other citizen.

Later that day, on Friday, November 8, I asked Shawn Browder, the Phoenix Policy Manager for Cognizant, about the Eric Ciaramella guidance. Shawn had engaged with Facebook via teleconference and in-person meetings on a regular basis for the last two years. This is what he said:

Ryan: Hey Shawn, dude like I've been getting a ton of those jobs this morning from the Ciaramella. I don't even know how to pronounce his name, but—

Shawn Browder: It's probably because uh, Facebook's classifiers are actively pulling the content into the queues.

Ryan: That's crazy, um yeah. I've been getting a lot of it. Gotta protect the guy though because he's, yeah its, his name is already out. But like it's gonna make it worse if they confirm that yeah he was the one.

Shawn Browder: And also like if it is confirmed that we're no longer gonna protect him under Coordinating Harm, still protect him under Violence and Incitement. But it's only while it hasn't been confirmed by government officials.

Ryan: Yeah we're getting a lot of it, and what else? Yeah I'm glad Skyler brought it up. What was the other thing that was trending? Naw, pretty much that.

Ryan: Is that, like with the name? Is that normal for the name? Cause just his name we delete that, the first and last name, that's pretty typical right, for—

Shawn: So, I need to pull up the guidance because, just wanna double check and make sure I'm not misinterpreting it. So, full name, first or last name if accompanied with a whistleblower claim, and any images accompanied with a whistleblower claim.

Ryan: Cool, I'll keep on sending those jobs over so we can track it.

Shawn: Yeah keep doing that so that I can raise them up to Facebook for visibility.

Ryan: Cool, thanks Shawn.

Shawn: Uh-huh, Thank you. [8]

Jobs showing Alexander Soros and calling him the whistleblower were also deleted. I specifically asked Shawn Browder on November 13, 2019. This is what he said about a specific post telling us to still delete even though we knew it was naming Alexander Soros when we were pretty sure the actual whistleblower was Eric Ciaramella:

Ryan: Hey Shawn, hey sorry to bother you. So even though they're showing Alexander Soros in that, I'll come back later if you're busy.

Shawn: So even though they're showing Alexander Soros, so, yes they're showing Alexander Soros, but they are making the claim that he is the whistleblower and showing his face, along with the name Eric Ciaramella, and so even though they're not showing Eric Ciaramella, we're still going to remove under that policy.

Ryan: Cause they're posting with the intention of saying that he is the whistleblower?

Shawn: Yup, of outing him.

Ryan: Okay thank you.

Shawn: No problem.[9]

Consider this situation. You are being accused of murder. The allegations are splashed all over the media. You can't do your job. And you can't even get any information about who has made the murder charge against you. We wouldn't let that happen to a regular citizen.

But this was what was allowed to happen to the president of the United States, in the middle of a campaign season.

Is there any objective person in America who can justify this chain of events?

It gets even worse. Facebook's initial post giving content moderators' guidance was deleted by their own bots. They were trying to cover their tracks. On November 19, 2019, I filmed a member of the policy team and asked about how the Facebook AI ended up deleting this internal post regarding Eric Ciaramella:

> [Hank Johnson‡ was on floor support and was answering a question I had about targeted cursing, and whether MF counted as motherfucker.]
>
> **Ryan:** Hey, so they brought back the uh, they brought back the post, the whistleblower post.
>
> **Hank:** Yeah because it was an automation that the SRT bot literally just, cause it's technically still Facebook and so it just automatically violates, it says the full name, remove. It did that multiple times.
>
> **Ryan:** So, they had the classifiers set up so that it'll automatically delete it, the bots will, the AI will.

---

‡ A pseudonym.

Hank: Yeah the AI will automatically try to, they're trying to [unwind?] through the content a little bit for his protection. Yeah it was remove, not by our accord. It was removed because of the bot.

Ryan: The post was removed?

Hank: Yeah.[10]

In February 2020, *Real Clear Politics* published an article with the ironic title "Can We Talk About Eric Ciaramella?" that highlighted what seemed to be some sort of cloak-and-dagger intelligence operation perpetrated by forces unknown against President Trump:

> Serious question: Are we allowed to talk about alleged Ukraine whistleblower Eric Ciaramella? It seems like so few are doing so even though he is one of the final missing pieces of the puzzle at the conclusion of the impeachment saga, a loose end that won't seem to go away.
>
> You can't talk about him on YouTube, as Senator Rand Paul learned.
>
> You can't talk about him on Facebook, as Ken LaCorte learned.
>
> Mainstream media, including Fox News, has a "Voldemort Rule" in place. Guests are told He Who Shall Not Be Named is anathema and cause for instant excommunication from cable news forever if his name is uttered.
>
> Twitter has remained Ciaramella-agnostic thus far, though some have reported there's an algorithmic suppression of Tweets that tag him.[11]

For those who think that an actual discussion is taking place about sensitive topics, rather than a managed discussion, it's shocking to realize that YouTube, Facebook, Twitter, AND Fox News were all suppressing information about Eric Ciaramella. The article continued as to why it was necessary to look into Ciaramella:

> The conversation needs to be had, as Ciaramella's involvement in questionable activities that extend back to before the 2016 election tells us he knows a lot more that needs to come to light. His fingerprints are all over Burisma, and not just as a whistleblower to the Zelensky phone call. Reports indicate he was engaged in covering for Hunter Biden while President Obama was still in the White House. His leaked conspiracy theory that Vladimir Putin ordered the firing of James Comey has never been fully resolved. Considering how much access he had to sensitive and classified White House information

through the NSC, CIA, and working for H.R. McMaster [Trump's National Security Advisor from 2017 to 2018], he must be questioned by the right people at some point in the very near future.[12]

Were we all being fooled by a mainstream media shell game? They were telling us we couldn't say the name, Eric Ciaramella, so obviously, we all wanted to say it. It's like being told you can't think of a pink elephant. You can't think of anything else but a pink elephant.

However, the mainstream media eventually told us the whistleblower wasn't Eric Ciaramella, but Lt. Colonel Alexander Vindman, who was once the Russia Political-Military Affairs Officer for the chairman of the Joint Chiefs of Staff, as well as being a military attaché at the US Embassy in Moscow. On August 1, 2020, Vindman wrote about his actions in an opinion piece for the *Washington Post*. Does he sound nonpolitical to you?

> A year ago, having served the nation in uniform in positions of critical importance, I was on the cusp of a career-topping promotion to colonel. A year ago, unknown to me, my concerns over the president's conduct and the president's efforts to undermine the very foundations of our democracy were precipitating tremors that would ultimately shake loose the façade of good governance and publicly expose the corruption of the Trump administration.
>
> At no point in my career or life have I felt our nation's values under greater threat and in more peril than at this moment. Our national government during the past few years has been more reminiscent of the authoritarian regime my family fled more than 40 years ago than the country I have devoted my life to serving.[13]

Let's be honest and say this guy sounds like the biggest drama queen to ever serve in uniform. No wonder they had to keep him under wraps, maybe even feeding us a couple other names like Alexander Soros or Eric Ciaramella, to throw us off the scent.

It seems like a game of cross and double cross, and like most people I'm still confused about who did what in the Ukrainian affair. I thought the Internet age was supposed to make the world more transparent, not allow powerful forces to come up with even more sophisticated ways to lie to us.

When I came out and accused Facebook of bias in June 2020, I put my name and face out there for the entire world to see. I didn't throw out phony clues or red herrings. The media was free to pick through my past and see if they could find anything to discredit me. But they couldn't find anything.

I'd worked for nearly two years at Cognizant, a company that didn't have much difficulty firing people if they didn't measure up.

The main criticism I seemed to get was that I was a white Mormon guy, which meant something must be wrong with me. Did Facebook try to protect my identity, as they did Eric Ciaramella's, when my Project Veritas video was released?

No, they did not.

And I wouldn't have wanted them to hide my name. I was voluntarily inserting myself into the public debate, or, as Facebook might say, I was engaging in "purposeful exposure." I take my responsibilities as a citizen very seriously. I'd tried to raise these issues first with my elected representatives, then with media outlets, and only after that didn't work, did I go to Project Veritas.

Facebook was making it clear for everybody with eyes to see.

There were the public figures they liked, who would receive every protection Facebook could offer.

For the public figures they didn't like, well, they faced the media equivalent of a ruthless drug cartel that only played by its own set of rules.

CHAPTER FIVE

Greta Thunberg—The Press Can Call Her Joan of Arc, but You Can't Criticize Her

In September 2019, Greta Thunberg was trending in national news stories. She gave a speech on September 23, 2019, in front of the United Nations in New York City about climate change.

Now remember, this is a sixteen-year-old girl. Think of the sixteen-year-olds you know and consider how well they understand life. It brings to mind the quote by Mark Twain that "When I was a boy of fourteen, my father was so ignorant I could hardly stand to have the old man around. But when I got to be twenty-one, I was astonished at how much the old man had learned."[1] At the time Greta gave her speech, she was two years past being fourteen and five years away from being twenty-one. Here are some excerpts from her speech to the United Nations:

> My message is that we'll be watching you. This is all wrong. I shouldn't be up here. I should be back in school on the other side of the ocean. Yet, you all come to us young people for hope. How dare you! You have stolen my dreams and my childhood with your empty words. Yet I'm one of the lucky ones. People are suffering. People are dying. Entire ecosystems are collapsing. We are in the beginning of a mass extinction, and all you can talk about is money and fairy tales of eternal economic growth! How dare you![2]

What I find remarkable is that the United Nations let a sixteen-year-old talk about climate change. We are told to listen only to those with expertise and training in a certain field. But Thunberg has none. Why are we being encouraged to listen to the experts in most fields, and yet in this instance we're told to listen to somebody who is clearly unqualified to evaluate the science and the issues? This is how her speech ended:

> You are failing us, but the young people are starting to understand your betrayal. The eyes of all future generations are upon you, and if you choose to fail us, I say we will never forgive you. We will not let you get away with this. Right here, right now, is where we draw the line. The world is waking up and change is coming, whether you like it or not. Thank you.[3]

I don't think I'm being critical of Greta when I say the typical sixteen-year-old is very passionate and also quite naive, has a streak of righteousness a mile wide, and can have tunnel vision when it comes to his or her most treasured beliefs. We've all been sixteen, and we're embarrassed by some of the things we thought, believed, and did at that time. Greta is no different.

The media was clearly using Greta for their own purposes, which was to create a figure who could criticize President Donald Trump, without being challenged. The perfect opportunity to create this battle came right after Greta gave her speech:

> When Thunberg was being acclaimed internationally in September for a passionate speech she gave about how her generation will feel the ravages of climate change, the president tweeted sarcastically: "She seems like a very happy young girl looking forward to a bright and wonderful future. So nice to see!" (At that same gathering, a United Nations climate summit in New York, Thunberg fixed a withering stare on Trump as he walked by.)[4]

Are you understanding the rules being laid down by the media? You can be sixteen years old, give the president of the United States a "withering stare," and yet that president can't even respond with a mild rebuke. Instead of asking the question of whether sixteen-year-olds should be in the national conversation, the media was busy turning Greta into a secular saint.

In December 2019, *TIME* magazine declared Greta Thunberg their "Person of the Year," accompanied by a picture of her standing on a rock at the edge of the ocean as a wave exploded in white foam, and underneath her name the caption read, "The Power of Youth." This is the image *TIME* wanted to paint of Greta:

> Greta Thunberg sits in silence in the cabin of the boat that will take her across
> the Atlantic Ocean. Inside, there's a cow skull hanging on the wall, a faded
> globe, a child's yellow raincoat. Outside, it's a tempest: rain pelts the boat, ice
> coats the decks, and the sea batters the vessel that will take this slight girl, her
> father, and a few companions from Virginia to Portugal. For a moment, it's
> as if Thunberg were the eye of a hurricane, a pool of resolve at the center of
> swirling chaos. In here, she speaks quietly. Out there, the entire natural world
> seems to amplify her small voice, screaming along with her.[5]

When I read the *TIME* article, I was personally embarrassed for the writers. It read like the opening of some overwrought young adult novel, you know, the kind where a young woman falls in love with a vampire, but a good and thoughtful one who only drinks the blood of animals.

The *TIME* article discussed how Thunberg got her start by skipping classes to protest climate change, then started camping out in front of the Swedish Parliament to further her campaign:

> In the 16 months since, she has addressed heads of state at the U.N., met
> with the Pope, sparred with the President of the United States and inspired
> 4 million people to join the global climate strike on September 20, 2019, in
> what was the largest climate demonstration in human history. Her image has
> been celebrated in murals and Halloween costumes, and her name has been
> attached to everything from bike shares to beetles. Margaret Atwood com-
> pared her to Joan of Arc. After noticing a hundredfold increase in its usage,
> lexicographers at Collins Dictionary named Thunberg's pioneering idea, *cli-
> mate strike,* the word of the year . . .[6]

Most sixteen-year-olds don't get to speak at the United Nations, meet the pope, and get to be compared to Joan of Arc. One could reasonably say she has had more effect on events than all but a few individuals living today. Does she need protection on social media? Does she lack the resources to defend herself? Let's ask the question about how she treats those with whom she disagrees. *TIME* gives us an answer:

> . . . She has offered a moral clarion call to those who are willing to act, and
> hurled shame on those who are not. She has persuaded leaders, from mayors

to Presidents, to make commitments where they had previously fumbled; after she spoke to Parliament and demonstrated with the British environmental group Extinction Rebellion, the U.K. passed a law requiring that the country eliminate its carbon footprint.[7]

You don't have to accept my view on how Thunberg deals with those who disagree with her. *TIME* gives it to you accurately: She provides a "moral clarion call to those who are willing to act" and hurls "shame" at those who don't share her views.

Greta is the media's fairy tale celebrity. She's a foreigner, and the media likes nothing more than a foreigner talking down to us when China is by the far the most egregious contributor to pollution in the world. In 2005, China surpassed the European Union and the United States for the most carbon dioxide emissions.[8] As of 2017, China produces nine billion metric tons versus five billion metric tons in the United States.[9]

Also, because Greta is a female minor from another country, any disagreement with her stance can be labeled by Facebook as racist, sexist, or xenophobic. Once again, even a disagreement on policy with Greta can be characterized as a petty attack on a minor child. Greta is not just a winner; she also gets double protections on social media!

I agree that minors should get additional protections on social media, and they do per Facebook's policy. However, their policy also differentiates between becoming involuntarily famous and voluntarily famous. Greta clearly falls within the latter.

With Greta being the perfect spokesperson for climate change who could not be attacked, what did Facebook do to protect her even further?

* * *

When we began getting the trending phrase "Gretarded" (Greta + retarded) directed toward Greta Thunberg, a number of policies came into play. Since Greta has autism (she suffers from Asperger's syndrome, a less severe form of autism), this could be considered hate speech because it could be perceived as an attack based on someone's PC (protected characteristic), which as of 2019 included autism. This is how *TIME* magazine described her condition:

Thunberg's Asperger's diagnosis helped explain why she had such a powerful reaction to learning about the climate crisis. Because she doesn't process information in the same way neurotypical people do, she could not compartmentalize that fact that her planet was in peril. "I see the world in black and white, and I don't like compromising," she told TIME during a school break earlier this year.[10]

Growing up as a teenager, I also saw things in black and white and treated things very literally. Although I was never diagnosed with anything, I had the ability to singularly focus on a topic and see it through to its conclusion. For Greta, I think she sees it as her mission to save the world from climate change. For her, there is no middle ground. This type of mindset can be helpful when dealing with mathematical dilemmas, but when it comes to global policies, it's important to be able to roll with the punches and see both sides of an issue.

The list of Protected Characteristics always included Down syndrome, but for the majority of my time at Facebook, autism was NOT a protected characteristic. Even if autism is a protected characteristic, I question whether Asperger's syndrome should be included, as people with Asperger's are generally able to function in the world, even though they may have problems in personal relationships. A joke in Big Tech is that the majority of the engineers have some degree of Asperger's syndrome, although accurate numbers are difficult to find.

But with the hate speech policy, it has to be clear that they're attacking her BECAUSE of her autism. If that connection isn't clear, then the hate speech policy doesn't apply. Unless the post mentions her autism and is calling her Gretarded, then the post would NOT be deleted for hate speech. It's very possible they could be attacking her policy position as retarded and using the play on words, but NOT attacking her based on her autism.

The next policy that could apply to this situation is the Bullying policy. If Greta were not a public figure, calling her "retarded" would be deleted, because she is a minor and this is a negative character claim, so that comment does not require a name/face match (NFM) to be deleted. If any other minor were called this on social media, that same minor would not need to report the comment for us to delete it. However, since Greta is a minor public figure, calling her "retarded" does NOT get deleted under the bullying policy.

If, however, an adult is called "retarded" on social media, that same adult would have to report it for us to take it down. On December 13, 2019, I recorded a conversation with a coworker named Paul Sernick,* who was a subject matter expert (SME), which is the equivalent of a supervisor.

---

* A pseudonym.

Ryan: How's it going my friend, how are you?

Paul: Going good, going good.

Ryan: Remind me of your name?

Paul: Paul.

Ryan: Paul, cool. There it is, on your name tag, Paul.

Ryan: Yeah we've been getting the Greta Thunberg again today. Um, so they reposted, I guess they're doing like an active for . . .

Paul: An exception, on her gretarded and retard.

Ryan: Well yeah they're doing an active pull of jobs so we're getting a lot more jobs. I was even getting jobs in like Swedish, so I guess they're . . .

Paul: I feel so bad for that girl. She's being used. It's so fucking clear, and no one is doing anything about it.[11]

The third policy that could apply here is the Cruel and Insensitive policy (C & I). This deletes or limits distribution of distasteful memes or jokes about people who died, victims of sexual assault, or people with disabilities. Once again, this would apply to Greta with her autism, but it would have to be clear. For example, if you had a meme showing a cartoon of a child with Down syndrome together with a caption that said, "This is Greta Thunberg," then the distribution would be limited on this meme.

If you had the same meme with an actual photo of Greta implying she has Down syndrome with a photo of another real person with Down syndrome, that would be deleted and removed under Cruel and Insensitive (C &I).

There are plenty of protections for Greta against attacks, even as a public figure. And minors are given additional protections, especially against any attacks that are sexual in nature.

However, Facebook went above and beyond their own written policy to scour the Internet for any mention of "Gretarded" and cleanse it of that phrase, which they considered to be abhorrent and offensive.

They proactively injected search terms to find the violating hashtags #gretard and #gretarded anywhere within their system. They told us to delete it under the Bullying & Harassment policy (B&H).

They made this decision based on the "spirit of the policy."

They even expanded the scope of these protections on October 2, 2019, to include the terms "retard" or "retarded" in addition to "Gretarded."

As you can see, this is an exception to Facebook's policy. Negative character claims against minor public figures are allowed, and in practice we allow these kinds of attacks against minors such as JoJo Siwa, the famous teen dancer, or Maddie Ziegler, another famous dancer from the reality show *Dance Moms*.

On Facebook, any minor public figure can be called stupid, retarded, dumb, idiotic, etc.

Facebook made a single policy exception and granted more protections for Greta Thunberg. She was espousing controversial climate views and garnering the world's attention.

But Facebook doesn't allow you to call her dumb.

Here are a few posts that give us more instruction regarding Ms. Thunberg:

Jessica Martinez[†]
October 1 [2019] at 3:20pm

#exception T54909092

TLDR SOP TO REMOVE GRETARDED
Action to Take on Job: Delete
Issue or Abuse Type: Bullying
Summary: Spirit of the Policy call to remove all instances of attacks aimed at Greta Thunberg using the term or hashtag, "Gretarded," EXCEPT for when used in a condemning context.

- Effective 10/2/2019: FB has expanded this exception to include the terms "retard" or "retarded" in addition to "Gretarded."

[Reaction by] You and 395 others   Seen by 814[12]

Shawn Browder to Cognizant CO Trending
19 hrs [ago]

† A pseudonym.

#trending #gretarded

T59077447

Facebook's US market team has identified potentially violating content containing the violating hashtags #gretard/#gretarded which will be proactively injected into [Post] North America. The classifier is also sweeping Workplace, and pulling in Workplace posts (e.g., 2596720943891662, 3274267025980885)

- Mon, Dec 23 – 200 jobs
- Tues, Dec 24 – 150 jobs
- Wed, Dec 25 – 100 jobs
- Thurs, Dec 26 – 100 jobs
- Fri, Dec 27 – 250 jobs
- Sat, Dec 28 – 300 jobs
- Sun, Dec 29 – 400 jobs
- Mon, Dec 30 – 368 jobs

Please reference the guidance found here.

Remember, standalone use of "gretard/gretarded" would NOT be violating, however if we have context to determine the attack is towards Greta Thunberg (i.e., a comment of "gretarded" on a picture of Greta, or re-sharing content containing Greta with the caption "gretard") we should **DELETE** for B & H.

[Reaction by] You and 227 others[13]

"Remove instances of attacks aimed at Greta Thunberg

SCALED POLICY EXCEPTION. Remove all instances of attacks aimed at Greta Thunberg using the terms or hashtag "Gretarded," "Retard," or "Retarded" EXCEPT when used in a condemning context.

Action to take	delete
Violation	bullying_and_harassment
Start Date	Dec 04, 2019
Exception ID	2728582903853671"[14]

"Hank

2 hrs [9/24/2019 approx. 9am]

#CruelAndInsensitive #GretaThunberg #Autistic #Autism #Aspergers
[ID] 938720363160645

[Description] Meme format depiction of Greta Thunberg that makes
fun of her for being autistic

[Action] Delete > Cruel and Insensitive > Serious disease/disability

[Reasoning] Greta Thunberg is publicly known to suffer from autism, and
the text overlay gives us further context that she is being mocked
for her disability. Since she is depicted in the content, the correct
action is to delete for C & I.

**Note: Autism is considered a serious disability, so be on the lookout for
Hate Speech violations. Additionally, she is a minor public figure and receives
certain protections under our Bullying and Harassment policy."[15]

In conclusion, we clearly see Facebook made a scaled policy exception to accommodate Greta Thunberg and prevent bullying toward her. I'm in favor of preventing bullying. But in this case, Facebook gave one-time enhanced protections to Greta and did not equally apply these protections to other minors. It appears Facebook was prejudiced in their decision to break their own rules to give Greta additional protections on their platform.

Regardless of who is being attacked and their policies, Facebook clearly has a double standard on how they treat public figures on the left side of the political spectrum versus those on the right. These public figures are discussing important political issues such as the environment, so it should be an equal playing field.

For example, the topic of climate change is very politically charged and can have a profound impact on our economy. Certainly, advocating to protect Mother Earth and our environment is a noble cause, and Greta has done good things for the world. Yet why are her ambassadorship and persona granted additional protections by Facebook? Why is Facebook bending over backward to protect her reputation?

The answer is simple: Facebook has become a bludgeon to enforce certain ideas and simultaneously attack so-called "wrongthink." In this case, calling a climate activist "retarded" is punished by Facebook, despite its not being against Facebook's own rules in the first place.

Greta Thunberg's identity is intrinsically linked to climate change, so by calling her "retarded," one could argue that it's simply an attack on her climate change ideology. We would not know who Greta is, if not for her stance on climate change.

Facebook clearly shows its own bias, and whether these decisions were made because of political pressure or social pressure does not matter. Facebook is acting as a publisher by enforcing one-sided rules and influencing political discourse on the platform.

The left and Facebook helped create Greta Thunberg into the perfect little dart to throw at President Trump that he would be powerless to defend himself against. It wouldn't be a fatal blow, but if they could get enough darts into him, they hoped it would be.

CHAPTER SIX

From Canada to Colombia, Facebook's Interference in Global Elections

During my nearly two-year tenure at Cognizant, I saw them place a huge priority on elections. As you read this chapter, keep in mind the following questions: Who authorized Facebook to actively monitor and track the US elections? Why is Facebook qualified to monitor and take action against political posts that contain election material? Why is Facebook enforcing election law in countries around the world? I'd also like you to think of what's called Hegelian dialectics. It means in convincing people to pursue a certain course of action, you first present a problem, then wait for the inevitable reaction, and finally provide the solution.

Consider former Facebook data scientist Sophie Zhang, who spoke out about what she witnessed of Facebook's role in influencing elections.[1] I find myself sympathizing with much of what she says, but the solution implicit in her complaint is deeply troubling and raises more questions than it answers. The gist of the article is that Facebook isn't doing enough to stop bots from influencing elections:

> The memo is a damning account of Facebook's failures. It's the story of Facebook abdicating responsibility for malign activities on its platform that could affect the political fate of nations outside the United States or Western Europe. It's also the story of a junior employee wielding extraordinary

moderation powers that affected millions of people without any real institutional support, and the personal torment that followed.[2]

Let's talk about what it really means when you say "moderating." It's the ability to ban a person or post. The way they're framing this issue, she's a do-gooder who wants to do the right thing, but Facebook isn't giving her enough support. Consider this lament she makes:

> I've found multiple blatant attempts by foreign national governments to abuse our platform on vast scales to mislead their own citizenry, and caused international news on multiple occasions. I have personally made decisions that affected national presidents without oversight, and taken action to enforce against so many prominent politicians that I've lost count.[3]

When I read what Zhang wrote, I found myself asking, "Who died and made you Queen of the Internet?" Did Zhang believe she had such powers of discernment that she could decide which information put out in a foreign country was true and which was false? However, the bigger issue is the fact that a role such as hers even exists at Facebook. The fact that she can make decisions affecting sovereign countries without oversight is shocking. Rather than being the Queen of the Internet, perhaps she's more like the overachieving supervillain Thanos from *Avengers: Endgame*, free from any restraints and able to wreak chaos on the universe, all under the guise of fixing things. If her actions sound implausible and cartoonish, they are. One person taking control of an entire country's political discourse sounds exactly like the plot of a villainous bad guy from my favorite superhero movie. Yet she was doing this on a daily basis on behalf of Facebook.

At another point, she wrote, "There was so much violating behavior worldwide that it was left to my personal assessment of which cases to further investigate, to file tasks, and escalate for prioritization afterwards."[4] My interpretation of Zhang's words was that she appointed herself the political vigilante for the Internet,* dispensing her own style of frontier justice where she was sheriff, judge, jury, and executioner. Her role was much larger in scope than mine, although we both deleted content. I could only delete one post at a time, while she could adjust and modify viral trends and themes on a global scale. And yet, for dispensing such unchecked power, she found

---

* Sophie Zhang's role as a data scientist was much larger in scope than my role as a content moderator. I was constantly supervised and graded on my actions, whereas she admits to having much more leeway to make "personal assessments" and take action "without oversight." I could only action one job at a time. I couldn't search for jobs to make decisions on.

herself "tormented." Yes, I imagine it can be very exhausting to be a dictator. So many decisions on who lives or dies, and who is allowed to speak and who cannot. The article continued:

> That power contrasted with what she said seemed to be a lack of desire from senior leadership to protect democratic processes in smaller countries. Facebook, Zhang said, prioritized regions including the US and Western Europe, and often only acted when she repeatedly pressed the issue publicly in comments on Workplace, the company's internal, employee-only message board.[5]

Still, she did not believe that the failures she observed during her two-and-a-half years at the company were the result of bad intent by Facebook's employees or leadership. It was a lack of resources, Zhang wrote, and the company's tendency to focus on global activity that posed public relations risks, as opposed to electoral or civic harm. In essence, this petty dictator was claiming she wasn't given enough power.

From my perspective, this is interesting because she's discussing how Facebook didn't want to protect democratic processes. Should Facebook be involved in protecting elections in the first place? What is the scope of their responsibility to foreign countries, or, for that matter, to the United States?

I also saw that Cognizant had the ability to help shape policy, but we did have to raise it up to the client (Facebook) on multiple occasions in order to get a response and for them to change their global policy. For example, in the Spanish market we saw large amounts of credit card fraud and noticed new code words that were not being deleted. We rose up these new terms, and after about eight months Facebook added these new "code words" to their list so we could delete them when we came across them.

However, I noticed that the 2018 US elections were a monumental priority for Facebook. In alignment with Sophie Zhang's experience, there didn't seem to be any apathy from upper management about US elections.

We had a special training session for the midterms, and we had to label certain types of content with the tag "VI," so that Facebook employees would then be able to take a second look. The policy that covers election fraud was called Coordinating Harm, and Facebook prohibited any attempts to mislead about election dates or pay for votes, etc. Oftentimes Facebook policy went above and beyond existing US election law.

Section 4 of the Coordinating Harm Policy reads:

4. Voter Interference

1. Voting interference through physical/verbal acts and misrepresentation of operational/procedural details include:

    1. Offers to buy or sell votes with cash or gifts, and methods for voting or voter registration.
    2. Statements that advocate, provide instructions, or show explicit intent to illegally participate in a voting process.
    3. Misrepresentation of the dates, locations, and times, and methods for voting or voter registration.
    4. Misrepresentation of who can vote, qualifications for voting, whether a vote will be counted, and what information and/or materials must be provided in order to vote.

2. Census interference through physical/verbal acts and misrepresentation of operational/procedural details include:

    1. Statements that advocate, provide instructions, or show explicit intent to illegally participate in a census processes' methods for voting or voter registration.[6]

As an interesting side note, several of Project Veritas's recent exposés portray violations of Facebook's policy relating to election fraud. On September 27, 2020, Project Veritas released a video showing an individual in Minnesota driving with multiple ballots inside his car.[7] A subsequent video showed an exchange of cash for ballots.[8] Another video released on October 27, 2020, shows political operative Raquel Rodriguez influencing someone to change their vote and also offering a gift to a voter.[9] So even by Facebook's standards, these individuals committed "voter interference." It's also important to note that Facebook wanted data and intel on election trends, not just election violations. We were asked to flag trends, and at one point they "urgently" needed visibility into discussion about the 2019 Democratic debates. Therefore, it wasn't just election law that Facebook was interested in looking into. We as content moderators were their eyes and ears, and without us they didn't know what people were talking about. In my opinion, our role was little different from that of informers in a totalitarian state. The term used for this information gathering is "civic insights requests."

Shawn Browder to Cognizant: North America Team

August 1 [2019]

[Reminder][Civic Insights Requests]

With the Democratic Debates the last two days, Facebook urgently needs visibility into any content that is coming through related to the debates. Please send any and all jobs relevant to the debates to COPHXTrends@cognizant. com for us to surface to Facebook.[10]

Another related post dealt with Trump's impeachment, which originated from the whistleblower mentioned in Chapter 4 of this book. We were told to flag any content that was trending related to the following:

Shawn Browder to Cognizant: North America Team

12/11/2019 4:05 PM

[HEADS UP] Impeachment Articles

Please continue to flag any content related to this market specific event that is trending, ambiguous, or may be a PR-fire risk to COPHXTrends@cognizant. com

House Democratic leaders have formally called for President Trump's removal from office on Tuesday, releasing two proposed articles of impeachment accusing him of:

- Abuse of Power: for pressuring Ukraine to assist him in his re-election campaign.
- Obstruction of Congress: for blocking testimony and refusing to provide documents in response to House subpoenas.

The House Judiciary Committee is expected to vote on the article of impeachment Thursday.

[liked by] You and 242 others [2 laughing reactions][11]

Additionally, in preparation for the 2018 US midterm elections, we were given weekly updates and recaps. Volume 6 of this weekly recap, which is dated September 11, 2018, states the following:

> Difficult Job Examples: "Monkeying things up," delete: "Hate Speech > Designated Dehumanizing Comparisons."[12]

This is in reference to Ron DeSantis's statement to Floridians saying they shouldn't "monkey this up" by electing gubernatorial candidate Andrew Gillum, which was interpreted by some as a racist dog whistle.[13] Another update given to us was the following:

United States 2018 Midterm Elections Reference Card

What are the party's priorities?

Democrats

- Gun reform is needed to keep schools safe (anti-gun laws)
- Disagree with the tax cuts for the rich while the poor continue to struggle
- Immigration (DACA) laws have been unsucessful, and we need to allow more immigrants into the country
- Need to gain power (win more votes) in the House/Senate on November 6th to be effective as a party

Republicans

- Trump's tax cuts created a strong economy and a strong stock market
- America has historically low unemployment
- [not visible] Fighting Isis
- Gun rights (pro-gun laws)[14]

Working at Facebook, I felt like I received a top-class education about civics, American politics, and election news. Another part of the "US 2018 Midterm Elections Weekly" update was that in Volume 8, they educated us about the importance of National Coming Out Day:

> Upcoming Event: National Coming Out Day, 10/11
>
> Summary: October 11th (10/11) is National Coming Out Day which is dedicated to awareness of civil rights for the Lesbian, Gay, Bisexual and Transgender community. We want to make sure that all reps are properly taking action on content. Trans Awareness week is also upcoming in mid-November so reps should be aware of this and potential violations as well.[15]

Yet another part of this "US 2018 Midterm Elections Weekly Recap," Volume 4, gave us education on the current status of the Mueller investigation:

> Mueller Investigation: the counsel investigating possible links between the Trump administration and Russian officials, ahs issues more than 100 criminal counts against individuals and three companies. Additionally, Michael D. Cohen a former lawyer for President Trump, pleaded guilty to charges that stemmed from evidence found by Mr. Mueller's inquiry, including campaign finance violations.[16]

Last, Volume 10 of this same midterm elections update, dated October 29, 2018, told us what kind of hate speech to delete about the caravan:

> Delete -Hate Speech
>
> Content in Review
> Published by Daniel Oy 11d 8h 17m
> Do it! https://www.foxnews.com/us/trump-threatens-to-call-us-military-to-close-southern-border-as-4000-strong-migrant-caravan-pushes-north"[17]

So apparently being in agreement with Trump's closing the southern border gets deleted for hate speech. Yet another post from October 22, 2018 gives us instructions about how to deal with hate speech with regard to Georgia allegedly purging its voter rolls of minorities:

[Heads Up] Georgia Candidate Improperly Purged 340,000 Mainly Minority Voters from Rolls

Summary:

Georgia secretary of state and Republican gubernatorial candidate Brian Kemp improperly purged more than 340,000 voters from the state's registration rolls, an investigation charges. 340,134 voters were removed from the rolls saying that they had moved, but they actually still live at the address where they are registered. Under Georgia procedures, registered voters who have not cast ballots for three years are sent a notice asking them to confirm if they still live at their address. If they don't return it, they are marked inactive. If they don't vote for two more general elections after that, they are removed from the rolls.

Potential Violations:

- Hate Speech against the minorities affected by the purge
- Coordinating Harm > Voter Fraud. Misinformation aroudn the voting dates and voting instructions

Support Articles: https://www.theguardian.com/us-news/2018/oct/19/georgia-governor-race-voter-suppression-brian-kemp[18]

Facebook's attitude toward US elections was anything but ambivalent. They stressed over and over the importance of having US content moderators to prevent the Russian election interference that happened in 2016. They told us this is why they hired so many new US content moderators after the 2016 presidential election.

Here is a transcript of a conversation I had on October 10, 2019, with Tom Akerman,[†] who was in a leadership role at Cognizant and at one point even visited Facebook's headquarters in Ireland. We had just finished a training session and were talking about election fraud:

Ryan: And now with Ukraine, like in 2016 there was a lot of election fraud. And now with Ukraine and this whole whistleblower thing, who knows if there's gonna be like more fraud. I wonder if Facebook's gonna be like clamping down on that kind of thing.

---

† A pseudonym.

> **Tom:** Oh I bet. They've been scrambling for the past couple of years to make sure that it's not gonna happen again.[19]

This conversation is somewhat unclear as to what the "it" is he's referring to. Facebook is trying to make sure *what* doesn't happen again? Is he referring to election fraud or Trump winning again? Or perhaps election fraud is Facebook's excuse for increased meddling in the election. At least from this conversation, it's clear Facebook placed a huge emphasis on election fraud on the basis of Russian influence in the 2016 presidential election, and many elements of this Russian collusion story have since been disproven.[20] To me, it seemed yet another clear example of a pattern and a mindset that ended up favoring the left at the expense of the right. For example, they clearly chose not to label Antifa as a criminal organization or a hate group, yet they wanted examples of violent nationalists.

On April 9, 2019, there was yet another internal post from Jessica Martinez. Jessica is originally from Ecuador and has a master's degree in public health. She is a Cognizant employee and is one of the leaders at Cognizant who interfaced with Facebook more than anyone else. She was one of the policy managers, and I interviewed with her for a job on the policy team in 2019. These internal posts are posted by Cognizant employees, but the wording comes from Facebook. In this post from April 9, 2019, the title is AD HOC REQUEST-POSSIBLE REAL WORLD HARM. And it continues discussing "Basque and Catalonian separatism/nationalism":

> In Spain there was a large movement from certain parts of Spain to secede. The national police were used to shut down this secessionist movement. It was a time of great turmoil for Spain, and is a still a sticking point in Spain politics to this day.
>
> [Summary] We are trying to get a representative idea of what type of content we see connected to Basque and Catalonian separatism/nationalism.
>
> Here is what we are looking for in terms of content:
>
> - Basque and Catalonian separatism/nationalism
> - Advocating for it or against it without specific reference to violence
> - Calls to violence, advocating for, encouraging violence in the name of said separatism/nationalism

[Ask] Please comment below with such examples.
https://www.newsroom.fb.com/news/2019/03/standing-against-hate/[21]
https://about.fb.com/news/2019/03/standing-against-hate/

It is very telling that Facebook wanted examples of advocating for or against nationalism even WITHOUT reference to violence. If the content doesn't violate Facebook policy, why do they care what people are talking about?

Also, taking this into context, this is a huge domestic issue that could influence the future of an entire country, in this case Spain. The leader of the separatist movement had to flee to Belgium at one point so he wouldn't get arrested.[22] But for Facebook, they included a link to their article about standing against hate. So is leading a nationalist movement hateful? Why is Facebook linking to an article about white nationalism in an article about secession and nationalism in Spain? Mind you, there is zero reference here to white nationalism. This is plain and simple internal politics and a part of the country wanting to secede.

Can you imagine similar involvement from Facebook when the United States wanted to break from Great Britain? Or when Texas had its own revolution against Mexico: whose side would have Facebook picked? I can imagine a similar post from 1835:

July 1st, 1835

AD HOC REQUEST—POSSIBLE REAL WORLD HARM

Texan nationalism and separatism

[Category] Ad hoc request
[Workflow] Events/Groups
[Market] USA/Texas
[Type] Coordinating Harm/Credible Violence

[Summary] We are trying to get a representative idea of what type of content we see connected to colonists in Texas and Tejanos attempting to gain independence from Mexico during the ongoing conflict.

Here is what we are looking for in terms of content:

• Content about President Santa Anna or Sam Houston

- Advocating for or against a Republic of Texas without specific reference to violence
- Calls to violence, advocating for, encouraging violence in the name of said separatism/nationalism
- Calls for violence against President Antonio López de Santa Anna

[Ask] Please comment below with such example
https://about.fb.com/news/1835/07/standing-against Texan-hate/‡

While at Facebook, I saw election training decks for the following countries: the United Kingdom, Poland, Hungary, Spain, Germany, Canada, Taiwan, Peru, Mexico, Argentina, and many more. In Canada, the Facebook election training deck includes descriptions of the main candidates for prime minister: Justin Trudeau, Andrew Scheer, and Jagmeet Singh.

Here we see the training deck has examples of market-specific trends. Facebook gave an example of candidate Jagmeet Singh targeted by hate speech with a dehumanizing comparison: "Never vote for a monkey."[23] There is no clear mention of his Sikh religion, but because he is wearing a head covering, it is assumed the attack is based on his religion, not his political candidacy. Wasn't it just as bad to call former President Trump "a pig"? And how about when we call lawyers "sharks"? Which animal comparisons will be allowed and which will be banned? Can you call somebody a "slug" without being unfair to invertebrates?

Also within this same election deck under market-specific trends is an example of violating hate speech toward immigrants. The photo showed a number of refugees in Canada standing next to a police officer. The caption states, "Look closely at this photograph. Do these look like refugees? Or are they opportunistic leeches coming to take advantage of Canadian 'kindness'?"[24]

This is also considered a delete for dehumanizing speech because it's comparing immigrants to an animal or insect (in this case, leeches). However, one interpretation of leeches is that they are parasites taking valuable resources from the system. In my opinion, there's a political discussion here insofar as that term refers to people abusing the system and taking from the legal citizens. Political debate requires vivid imagery to drive home arguments across all aspects of the political spectrum. We may agree with those images, we may be revolted by them, but they have been part of

---

‡ Fictional post created by the author to characterize what Facebook's stance may have been if social media had existed in the nineteenth century.

our political culture for hundreds of years. You can find political cartoons from the 1860s depicting President Lincoln as a large gorilla because of his exceptional height and long arms. We may find ourselves offended by such depictions, but the decision for most of our history has been to allow these images, and let the public decide whether a good argument is being made by their use. If the image were to call them animals, that's a little different. But here they're using the word "leech," which has a political meaning for someone who drains the system of resources. There are perhaps Canadians who have compassion for refugees but are using the term *leech* to describe the behavior of the refugees, not the refugees themselves. However, Facebook has banned this type of political discourse. Facebook also wanted their own employees to review certain types of election content in Canada. It wasn't good enough for contractors to review the content; they wanted their own people to review such important information as:

What should be escalated: [to Facebook employees]

- Trends and viral posts that are related to elections including humorous & satire posts.
- Hate speech or bullying & harassment related to elections
- Threats to political candidates (even if they don't meet the threshold for policy removal)
- Attempts for voter fraud and spreading misinformation around the election—please take action according to policy- and send us the SRT ID on the tribe
- Potentially privacy violations related to elections

How to escalate

- Please follow normal procedures for escalating content in the MSP policy workflow tribe and the NA-OS workflow and policy support tribe.
- Please tag Jessie Banks and Celina Vicuna for any Canadian-related escalations.[25]

This is why former Facebook data scientist Sophie Zhang's statement about Facebook's placing emphasis on North American elections coincides with my own observations. Facebook clearly had enough resources for their own employees to review humorous and satirical posts about Canadian elections.

My favorite screenshot of the same training deck is titled "Expected Violations."

In other words, "what we hope to delete" or "what we plan on deleting." It reminds me of the 2002 movie *Minority Report,* where the PreCrime task force uses PreCogs to see the future and then arrests people before they commit a crime. It's also important to remember that these instructions are being given to content moderators who are being paid a little more than twenty-eight thousand dollars a year and know they can be fired if their "accuracy score" falls below 95 percent for a couple of weeks. So they have plenty of incentive to fall in line and to not raise objections.

Expected Violation Types

- Voter fraud/suppression
- Credible violence (vulnerable person)
- Harassment
- Impersonation
- Hate speech
- Fake accounts

*This is NOT a final list of possible violations or issues; many more possibilities can be found during this period.[26]

In addition to Canada, Facebook also influenced a revolution in Venezuela;

[Jessica Martinez] to Cognizant CO ESLA High-Pri

April 30 [2019]

#HEADSUP –[VENEZUELA] EXCEPTION FOR VIDEO OF COLECTIVO LEADER

[PRFireRisk] Yes
[Workflow] IG Video, CM
[Market] VECAM, ESLA
[Type] Violence and incitement

[**Summary**] There is a video in which Valentin Santana, leader of the colectivo "La Piedrita" makes a call to "defend the revolution with arms" and go to Miraflores.
[**Action**] Please

- **Delete** all instances of the video (depicted below) with neutral caption, no caption, or caption in support of Valentin Santana and/or his message.
- **Ignore** instances of the video with condemning caption or context.

[Sample] 976658446066579

Let us know if you have any questions. Thank you![27]

This post is following Facebook policy by deleting posts that advocate for violence. Valentin Santana is a pro-Nicolas Maduro figure, and the revolution being referred to is the Bolivarian Revolution, which commenced with Hugo Chavez.[28] Valentin Santana is the leader of a "colectivo," which is an armed community support group for the Chavez/Maduro regime.[29]

Regardless of whether you support the Maduro regime or not, this shows Facebook's ability to quash an armed "revolution" in a foreign country.

* * *

Did you know that Facebook tracks political parties as if they were a consumer product? It's not democracy versus communism; it's Brand X versus Brand Y, it's Coke versus Pepsi. They even give it a cool Madison Avenue marketing term called "Brand Safety."

I believe Brand Safety labels content for advertising purposes, and to allow customers to choose what kind of ads appear on their groups or pages. I can't help but wonder what kind of "Brand Safety" ranking Facebook would have given our Founding Fathers. Here's what Facebook was doing in Latin America:

Jeff Franklin§ to [CTS] Brand Safety Review Team – São Paulo

October 1 [2019] at 9:18am

---
§ A pseudonym.

[Spanish] Hello! Since we consider the political parties as political affiliation, I found it interesting to share with you the main ones of Argentina, Mexico, Dominican Republic, and Spain, since these nations have greater focus on this kind of discussion. These are:

Mexico Mexico
National Action Party (bread)
Institutional Revolutionary Party (Pri)
Game of the Democratic Revolution (Prd)
Work Game (PT)
Green Green party of Mexico (Pvem, green)
Citizen Movement (MC)
National Regeneration Movement (Brunette)
Argentina Argentina
Civic Coalition Ari
Revolutionary Army of the people
Front of the left and the workers
In front for victory
Renewal front
Free movement of the south
Socialist Movement of workers
Communist Party of Argentina (Extraordinary Congress)
Revolutionary Communist party of Argentina[30]

The fact that Facebook was hiring Brazilians to track political parties should be of concern to Brazil and to the rest of the world. If Facebook does not censor political speech, then why does it need a list of political movements across Latin America? I can't help but wonder if these are test cases for a more global system of control over what we see and, as a result, are allowed to think.

Should it come as any surprise that "Brand Integrity" has also been rolled out in the United States?

* * *

In the United States, we see a similar categorization of concepts and whether Facebook considers them political. The following post is from the Florida Brand Integrity (BI) team. This team was housed in Tampa, Florida, and consisted of Cognizant employees. This team would also label content and determine whether or not it was political. The whole purpose of this department is to determine monetization on the content. The fact that Facebook monetizes or demonetizes content based on whether it's a political/social issue means that they are tracking political speech and can limit content based on its political message.

Tiffany Peters⁵ to Florida BI Team
June 4, 2018

"Anything, anywhere outside of North America is NOT potentially a political/social issue unless America is mentioned (Ex. Trump gains Putin as an ally would be marked as potentially political/social vs China's economic status, would not be potentially a political/social issue).

Organizations such as ASPCA, PETA, Breast Cancer Awareness Groups, Autism awareness, Planned Parenthood, etc., are usually (based on the ad) potentially political/social issues, since they are advocating for awareness of that issue on a social/political issues stance via the policy.

The American flag is NOT potentially political/social UNLESS it is being sold (ex. An American flag being sold, a shirt with the American flag on it would be yes to potentially political/social vs a flag outside of a house up for auction, Memorial day sales, etc.).

Groups such as Black Lives Matter and Blue Lives Matter are not currently considered a political/social issue, therefore it will be marked as NOT potentially political/social issue. (Regarding Blue Lives Matter, in cases of where the American flag with a blue line through it is being sold/promoted on a shirt, watch, ad, etc., it is not currently known for sure whether it is a potential political/social issue or not.)"[31]

We see that Facebook has a keen interest in the election process in certain countries, while deprioritizing other less-important countries. Their focus goes beyond the scope of a country's own election law and expands into conversations, trends, humor, and satire related to elections. In my opinion, these actions are clearly attempts to foist themselves atop the election process and influence elections. This should scare elected leaders throughout the world, knowing that a company with such utter disregard for privacy

¶ A pseudonym.

and local laws is tracking election trends and its own unique rules to enforce election law. The amount of hubris Facebook assumes in deciding it knows best, not just for the United States but for the world, is truly astounding. When a corporation is deciding the conversations that are allowed around the globe, we need to have more than a "conversation." This is a global theft of our rights and demands the equivalent of an international SWAT team knocking on their door.

CHAPTER SEVEN

Can You Love Your Country without Being a Racist?

Facebook is a global company and has a staggering 2.7 billion monthly active users.[1] As such, it's normal that within the platform there are many political discussions happening in countries all around the world. Many political campaigns and politicians use Facebook as a way to reach out and connect with voters. The Facebook website for government, politics, and advocacy states the following:

> Facebook has made investments in teams and technologies to better secure elections. Since 2016 we've tripled the size of our teams working on safety and security issues to include more than 35,000 people, and we've created rapid response centers. We've also made significant improvements to reduce the spread of misinformation and provide more transparency around ads about social issues, elections or politics. Elections have changed and so has Facebook.[2]

But how much of political speech is censored or considered "misinformation" by Facebook, and what role does a sovereign country play in dictating the types of conversations happening on social media?

If a dictatorship were able to use Facebook to suppress types of political speech, wouldn't that be a great tool for silencing political dissent?[3]

In the Buzzfeed article referenced in the previous chapter, we discussed a memo from a recently fired Facebook employee. This employee's opinion is exactly that: heads of state in Latin America and other parts of the world use Facebook to control elections and mislead their citizenry.

Data scientist Sophie Zhang said, ". . . ultimately I was the one who made the decision not to push more or prioritize further in each case, and I know that I have blood on my hands by now."[4]

The basic gist of the article is that Facebook is not doing enough against fake accounts, and politicians in power are gaming the system to control discourse on the social media platform.

Sophie Zhang was able to investigate what she believed was fake and fraudulent activity. She claimed her ability to influence elections and the democratic process was immense. But she felt she wasn't able to do enough and found rampant trolling and inauthentic behavior influencing the democratic process throughout the world. Once again, Sophie is NOT a political scientist or government employee. She is a data scientist, and I find it odd that she's responsible for controlling the election process in foreign countries as a Facebook employee.

Zhang mentions an example in Honduras where many fake accounts were created to boost or spread content benefiting Honduran President Juan Orlando Hernàndez. It took her a year to take down these fake accounts, but they were brought back rapidly after Facebook's takedown operation.

Zhang stated that Facebook "didn't care enough to stop" and "they have bigger fish to fry."[5]

* * *

What I find curious is that Facebook didn't have the resources to prioritize certain countries, but they were able to prioritize the deletion of nationalist movements in Spain, Hungary, and Poland. Perhaps this is what she was referring to when she said they have bigger fish to fry. It's likely the "bigger fish" for Facebook are the nationalist movements, the US elections, and what they call right-wing hate figures.

In the October 28, 2020, Senate Hearing, Mark Zuckerberg testified that they spend billions on content moderation:

> Senator, we have more than 35,000 people who work on content and safety review. And I believe our budget is multiple billions of dollars a year on this.
> I think upwards of three or maybe even more billion dollars a year, which is a greater amount that we're spending on this than the whole revenue of our company was the year before we filed to go public in 2012.[6]

Sophie Zhang stated that she raised the impact Facebook's inaction was having on elections and was told that "human resources are limited." If human resources are limited, why did Facebook hire thousands of content moderators for the US elections? This shows that Facebook prioritizes certain elections over others. Clearly, human resources are not limited when it comes to the United States and its civic matters.

The examples of election interference in the Buzzfeed article were selected from a 6,600-word memo, and many of the examples in Latin America show Facebook indirectly favoring US-supported politicians such as Juan Orlando Hernández from Honduras, and Facebook allowing inauthentic activity which favored the opposition candidate who was running against President Evo Morales of Bolivia. Both of these examples show Facebook indirectly supporting Honduran President Juan Hernández, a US ally, and acting against Evo Morales, a president attempting to become a dictator in Bolivia.

It's unclear whether Facebook was purposely favoring politicians in Latin America who are friendly to the United States. The United States does have a long history of influencing elections in Latin America, so I wouldn't put it past the United States to have the ability to influence Facebook's decisions on foreign soil.

* * *

During my time at Facebook, the global policy team clearly prioritized hate speech and groups associated with it, as defined by the Anti-Defamation League(ADL) and Southern Poverty Law Center (SPLC). For example, in 2017, Facebook acted on the SPLC's recommendations and deleted fifty-seven groups from a list of two hundred suspect groups.[7]

The SPLC has a long history of mislabeling pro-life and conservative entities as hate groups, including the Family Research Council.[8] This labeling led to an armed mass murder attempt against the Family Research Council.[9]

This idea of prioritizing makes sense based on one of Ms. Zhang's excerpts. The Buzzfeed article states:

> But she also remarked on Facebook's habit of prioritizing public relations over real-world problems. "It's an open secret within the civic integrity space that Facebook's short-term decisions are largely motivated by PR (public relations) and the potential for negative attention," she wrote, noting that she was told

directly at a 2020 summit that anything published in the New York Times or
Washington Post would obtain elevated priority.[10]

Obviously, if Facebook were to allow a Proud Boys page to flourish, this
would receive a great deal more media attention than if an Antifa page were
left up. The cancel culture forces on the left are far more powerful than
those on the right, and the leftist culture is prevalent throughout the news-
rooms of the *New York Times* and the *Washington Post*.

This brings into question Facebook's role as a publisher, if indeed they
are using the *New York Times* and the *Washington Post* as figurative canaries
in the coal mine to detect potential public-relations fires. If they are prioritizing certain news outlets as a basis for business decisions, then they may be
inclined to prioritize those same news outlets in their news feed.

I admire certain aspects of the courage of Sophie Zhang, although some
of her solutions I find terrifying. Her perspectives confirm my belief that
Facebook can influence global elections, whether intentionally or through
acts of omission. Ms. Zhang clearly showed Facebook's apathy to obvious
trolling and manipulation through fake accounts in many democracies
throughout the world.

I directly witnessed Facebook prioritizing the US elections in both 2018
and 2020, and according to Sophie Zhang, they let other countries fall by
the wayside.

However, they still applied their leftist policy in Latin America and
targeted nationalist movements throughout the world.

The question posed by Facebook's apparent dislike of nationalist movements is "With what do they seek to replace it?" If Facebook does not approve
of people showing loyalty to their own country, to what does Facebook wish
them to pledge their support?

* * *

Does Facebook tailor each policy specific to a country and conform to that
country's election laws? How is the discussion of immigration law enforced
on social media?

In this chapter, I'll focus on nationalism, which has been a main focus
of Facebook's censorship campaign.

Facebook has outright banned many nationalist groups throughout Europe. Mind you, these are not racist groups, just individuals who
believe in putting country first. Even in Poland, Facebook banned their

Independence Day March along with a few very large nationalist groups, with hundreds of thousands of followers.[11] This first took place in 2016, and this kind of censorship has continued to the current time.

Many Facebook groups in Poland, Hungary, and Spain were against Muslim immigration into their countries, and political groups such as these were censored or deleted on Facebook.[12] The hate speech policy lends itself to disallowing any discussion against immigration, especially when religion is mentioned. For example, if I simply say, "Keep Canadians out of the United States," this would be deleted for hate speech, because I'm excluding people based on nationality.

* * *

Spain also experienced a purge of right-wing nationalist groups right before their National Election.[13]

I did a virtual interview in Spain in July 2020 with a news outlet called El Toro TV, on a specific program called *El Gato El Agua*.[14] In this interview, we talked about the extreme censorship right-wing groups are experiencing in Spain.

Another political party in Spain that has experienced censorship is the Vox party (not to be confused with the news outlet in the United States). The Vox party in Spain is against political correctness, advocates for limited government, and supports family values. They have grown exponentially in the last few years. Vox is not a nationalist party. But as a right-wing party, they have experienced a great deal of censorship. I spoke with the secretary of the Vox party, Gabriel Ariza, around June 30, 2020. He wanted to know if I had evidence of Facebook directly censoring the Vox party. I did not. However, I do have a post that was given to content moderators, warning us about violence associated with Basque and Catalonian Nationalism/Separatism.

There was a large movement from certain parts of Spain to secede. The national police was used to shut down this secessionist movement. It was a time of great turmoil for Spain and is a still a sticking point in Spanish politics to this day.

Here is the guidance Facebook gave us, which I also shared in Chapter 6:

April 9th 2019,

AD HOC REQUEST-POSSIBLE REAL WORLD HARM.

[Summary] We are trying to get a representative idea of what type of content we see connected to Basque and Catalonian separatism/nationalism.

Here is what we are looking for in terms of content:

- Content about Basque and Catalonian separatism/nationalism
- Advocating for it or against it without specific reference to violence
- Calls to violence, advocating for, encouraging violence in the name of said separatism/nationalism

[Ask] Please comment below with such examples.
https://www.newsroom.fb.com/news/2019/03/standing-against-hate/"[15]
[Updated useable link: https://about.fb.com/news/2019/03/standing-against-hate/]

A few Spaniards reached out to me after my interview in July and expressed their discontent about the situation in Spain. Here is what one of them wrote to me, translated into English:

On censorship: it's necessary to publish information that is purposefully hidden by the government's media arm and journalists who go along with the narrative. By publishing this new information, we are able to restore justice in small way and restore some balance, while fighting against absolute totalitarianism and the corrupt elites' abuse of power.

Here is some of the info that flies in the face of the controlled narrative and lies and exposes the abuse of power. She went on to explain in detail much of the corruption that is going on and how many public officials are being arrested. But it's nowhere to be found in the news. She continues, saying, "My friends, this brutal and controlled media narrative and extreme censorship that exists today in Spain is the precursor to totalitarianism. The best way to fight against this is to share the truth as much as possible."[16]

Facebook was very proactive in Spanish politics. Another post in April 2019 said the following:

[Category] Ad hoc request

[Workflow] Any

[Market] Spain

[Type] Hate Speech, Bullying & Harassment, Coordinating Harm/ Publicizing crime, Credible Violence

[Summary] The Spanish general election is set to take place on April 26, after weeks of campaigns. The only presidential debate ahead of elections will be televised on April 23.

https://www.euronews.com/2019/04/12/spanish-general-election-2019-who-are-the-candidates-and-what-are-their-manifestos

Please flag edge cases that could lead to potential community risk or real world harm[17]

A similar post dated April 15, 2019, says the following:

Jessica Martinez shared a link to the group: Cognizant CO ESLA High-Pri

April 15

Spain Elections-Ad Hoc Request

[Category] Ad hoc request

[Workflow] All

[Market] ESLA/Spain

[Type] Graphic Violence

[Summary] Police in Spain's northern city of Bilbao have clashed with hundreds of protesters outside a rally by a far-right party that is running Spain's general election this month.

[Ask] Please comment below with such examples.[18]

* * *

The world is a dangerous place, and we often see attacks and mass murders throughout the world. Throughout the Middle East, Christians are persecuted and murdered,[19] and we also see similar persecution of Muslim Uyghurs in China.

As a content moderator, I saw a vast variety of violence. I saw humans being tortured by Mexican cartels and watched as awake and coherent individuals had their arms sawed off by knives and other forms of horrific torture. I saw Al-Qaeda beheading "infidels," and I also saw pornography, snuff videos, animal abuse, and child abuse.

When we see mass murders like Columbine, the Las Vegas shooting, or others, how do we compare these tragic events to other organizations and groups? Who is worse? Which group is more brutal? Who should be prioritized by Facebook? Do we look at the hard numbers of deaths, or do we examine motives and evaluate which ideology is more "dangerous"?

In one horrific attack in New Zealand in March 2019, now known as the Christchurch Mosque Shooting, Brenton Tarrant murdered fifty-one Muslim worshippers inside a mosque.[20]

Any rational human being denounces this senseless violence in the strongest form possible. However, I did see that Facebook treated this mass murder differently from other attacks or murders. For example, in 2020 alone, there were hundreds of attacks and more than ten thousand murders of Christians in Africa and other parts of the Middle East.[21]

Right after the New Zealand attack, Facebook placed severe restrictions on how the attack could be shared. For a time, we banned completely any video footage of the incident, which was going viral on the Internet. In the aftermath of the attack, Facebook created a new section of their policy that made white nationalism a "designated hate organization" instead of just a "hate organization." For Facebook, white nationalism is more dangerous than any other type of racism that was already banned on their platform.[22]

Whenever white people do something bad, Facebook is quick to alter its policy and immediately inform its thousands of content moderators. Also, as of December 2020, Facebook has now codified this deprioritization of attacks on white people.[23] And with regard to nationalism, it's important to note that the nationalism by itself is not inherently racist, yet in the media if you hear the word "nationalism," it is now inextricably tied to "racism." Nationalism is defined as "devotion, especially excessive or undiscriminating devotion, to the interests or culture of a particular nation-state."

Here is a section of the Dangerous Individuals and Organizations policy that I took a video of on 10/12/2019 around 11:00 a.m. and transcribed, which discusses Facebook's training content regarding "Hate Entities, Ideologies and Crimes":

Evaluate

Thought process for evaluating content expressing White Supremacy, Nationalism, Separatism, and Nazi/Fascist Ideologies:

Linking this training content to Dangerous Orgs policy & process language

IS [implementation standards] #6   KQ[known questions] L

Why does this policy language exist?

Prior to 3 April 2019, Facebook allowed PSR [Praise, Support, Represent] of all forms of nationalism and separatism, which includes white nationalism and white separatism hate ideologies. By continuing the allowance of content expressing these hate ideologies is inconsistent with how Facebook treats other hateful ideologies such as Nazism, in addition to note accounting for the historical context and current use of these terms.

In order to maintain consistency, and better protect our community of users from the potential of [illegible] harm, Facebook expanded upon existing hate ideology policy language, as well as moved the current hate ideologies from Hate Speech policy language to the Dangerous Individuals & Organizations policy language.[24]

We see here how any mention of nationalism is considered racist. In the below guidance, there is zero mention of white nationalism; instead, we see a protest organized by nationalists. On April 29, 2019, there was a protest at a bookstore in Washington, DC, and I received the following guidance:

#Headsup #HateOrg #bookstore

A group of self-described nationalists interrupted an event Saturday at a Washington D.C. bookstore with an author of a book titled "Dying of

Whiteness: How the Politics of Racial Resentment is Killing America's
Heartland"

> Policy Guidance: Any praise representing or in support of this event will be
> actioned for PRS of a Hate Org[25]

Nowhere in this post is the group protesting identified as white nationalists. It clearly labels them as nationalists, but NOT as white nationalists. Now with the context, we know they are a protesting a book called
*Dying of Whiteness: How the Politics of Racial Resentment is Killing America's
Heartland.*

Is it so unreasonable to ask whether there are patriotic Americans who
disagree with the characterization that whiteness is killing America? Yes,
and it may be a group of Americans (nationalists) who are protesting this
attack on race.

Can you imagine if there were a book titled *Dying of Blackness*, and do
you think there would be an issue if black people showed up to protest a
book implying blacks are killing America?

If there were truly racist groups at this protest, then Facebook might
have some rationale for their decision. But this post is not clear about who
was in attendance. And last I checked, even racist groups can show up and
protest outside of a bookstore. I searched for the book online (*Dying of
Whiteness*) and found the following summary:

> In the era of Donald Trump, many lower- and middle-class white Americans
> are drawn to politicians who pledge to make their lives great again. But
> as *Dying of Whiteness* shows, the policies that result actually place white
> Americans at ever-greater risk of sickness and death.
>
> Physician Jonathan M. Metzl's quest to understand the health impli-
> cations of "backlash governance" leads him across America's heartland.
> Interviewing a range of everyday Americans, he examines how racial resent-
> ment has fueled progun laws in Missouri, resistance to the Affordable Care
> Act in Tennessee, and cuts to schools and social services in Kansas. And he
> shows these policies' costs: increasing deaths by gun suicide, falling life expec-
> tancies, and rising dropout rates. White Americans, Metzl argues, must reject
> the racial hierarchies that promise to aid them but in fact lead our nation to
> demise.[26]

This book attempts to plant racial resentment as a motivating factor in the support of conservative policies such as progun laws or decentralized healthcare. It argues that White Americans must reject the racial hierarchies that lead to these policies that subsequently negatively affect one's health and lead to death and higher dropout rates.

In other words, whites who oppose big government policies will get sick and die sooner because they're white, racist, and support Trump. Granted I haven't read this book, so it may be more nuanced than I make it out to be, but the main message is that white Americans are placing their health in peril by following and supporting conservative policies.

Imagine another book discussing how blacks who oppose government healthcare and government control in their lives are going to die sooner because they're black and support Obama. I don't think that would go over very well with the black community.

Therefore, I'm not surprised that a few white people showed up to protest a book saying the same thing about the white community. Are you?

* * *

Below is transcription and notes from video I took of Facebook's election training decks. As you can see, they provide training for many countries throughout the world. I had full access to these training decks as a US-based content moderator and did not have to gain unauthorized access or hack into anything to find this material. The following are notes I took based on video I filmed regarding these election training decks and a few other documents:

9/12/2019 Thursday AM Video

Transcript & Notes

Video 1
7:33 dangerous individuals and organizations policy
8:28 hate figure list: Faith Goldy, Gavin McInnes, Hitler, Nazi leadership

Video 2
0:01 Tommy Robinson on hate figure list
1:05 From Dangerous Orgs policy, click on Market specific content, which brings us to a screen

Which lists many different countries and their specific policies on hate speech and also election information.

3:07 start going through each country's deck with info about elections
> Countries/Market:
> Afghan
>> Afghan Elections—updated July 29, 2019
>> Shows images of candidates, along with key players, political parties, and journalists and activists.
>> Raise any trending examples to Facebook staff
> Albania
>> Hate Speech Market clarifications—nothing specifically on elections
> Austria
>> Austria Elections 2019—updated sept 10, 2019
>> Number of voters, voting breakdown from last election, current prognosis
>> Prominent candidates
>> 7:06 minute mark—candidates—includes the link to their Facebook & Instagram pages

> Arabic
> Bulgarian
> Burmese See video 3 at 3:40
> CJK
> CIS
> Czech & Slovak
> ESLA (Latin America)
> Francophone
> Germany

Video 3
1:32 Presidential Election Primer 2020

3:40 Burmese Hate Figures
> Wirathu—Myanmar

9:19 Japanese Hate speech
9:28 Chinese Hate Speech

Video 4

3:37 CIS Market Dangerous Orgs

3:34 Why is Ukraine's National Guard, the AZOV battalion classified as a hate org?

3:41 LDPR

https://www.stalkerzone.org/why-ukrainian-soldiers-desert-and-switch-to-the-ldprs-side/

4:20 Czech & Slovak Market insights

6:45 German Elections

Also 7:01 voting statistics, how old to vote, ways to vote, voting systems

List of candidates for each party—over 72 candidates

7:50 Regional Elections—Brandenburg & Saxony 2019

9:10 India General Elections 2020—updated 4/4/2019

Video 5

0:03 Indonesia Hate speech guidelines

0:09 Israeli Re-Election 2019—updated August 16th 2019

Parties & candidates, party leaders

Until 38 second mark

Then again from

0:54 more Israel election PowerPoint

1:07 Bentzi Gopstein is a hate figure and is banned from Facebook

https://www.lehava-us.com/interview-with-lahava-ceo-bentzi-gopstein/

1:43 Israel elections—what to escalate

3:14 Kazakhstan 101 map of Kazakhstan, general info about demographics

Prominent public figures, pie chart of the market share of Kazakhstan

4:11 Kurdish CT Training deck

4:27 Poland Independence Day October 16, 2018

5:12 Polish Independence day deck

6:15-7:00 targeting Polish Independence Day where LGBTQ flag was burned

9:29 Polish Dangerous orgs

9:43 PSR examples of Praise, Support, Represent of dangerous orgs in Poland

Video 6

1:11 More PR fire Examples Polish Independence Day

2:22 Finish, Polish Independence Day slides

3:23 Thai Geoblock Guidelines can't criticize monarchy

4:28 Turkish Hate speech, bullying, etc.

4:36 Western Balkans, Hate speech, dangerous orgs, misrepresentation
(privacy violations)

6:44 Vaccine Misinformation Slides

9:00 Examples of non-violating vaccine information vs violating

Video 7

Anti-Vaccine and how they control the hashtags—what's allowed what's not
allowed[27]

In addition to Facebook having a global reach and influence in political and
social movements worldwide, they now want to control what you can read
about vaccines. Does Mark Zuckerberg now have a medical degree?

In closing, this final guidance portrays Facebook's attempt to lump
together right-wing groups (nationalists, separatists) as dangerous and
fringe:

**Shawn Browder posted Regional Far-Right Extremism to Cognizant:
North America Team**

September 12, 2018

[Market Insights Request] + Regional Far-right Extremism

Type: [Far-right Extremism + Terrorism]

Overview: Facebook is beginning to investigate the presence of far-right
extremist and extremist terrorist groups (outside of the US) on
Facebook and Instagram, and need to understand what content reviewers
for the North America market are seeing at the ground level.

In order to mitigate risk for Facebook, we are collecting job IDs that
highlight trending content related to regional extremist groups and
individuals, and job IDs for content that highlight gaps in the policies
and guidelines surrounding these individuals and orgs.

Please collect and surface examples on an ongoing basis and send to both
Suzannah
Fischer and myself."

[reaction] You and 464 others[28]

The most fascinating part of this guidance is that it explains that its purpose is to highlight gaps in the policies regarding these groups and individuals. This shows that Facebook is hyperfocused on one type of extremism and cannot let anyone fall through the cracks that shares dangerous right-wing ideas.

Yet its myopic vision allows it to unintentionally or intentionally ignore mass atrocities committed against Christians in the Middle East, and also allows them to continually ignore mass violence and insurrections committed by Antifa or Black Lives Matter. And have I provided you with any examples of Facebook going after China? No, because apparently, Facebook has no problems with the Chinese communist party.

Facebook truly considers nationalist groups as enemy number one. They are persistently targeting these groups worldwide, which in effect is an assault on the sovereignty of every country in the world.

What are we going to do about it?

CHAPTER EIGHT

Orange Man Bad: Facebook's Guidance to Content Moderators about Trump Statements

Trump was vilified by the media as a racist, and this sentiment was echoed by Facebook's hate speech policy. This mischaracterization of Trump also extended to Trump supporters. In this chapter, I will present concrete examples of how Facebook's policies favored liberals and disfavored conservatives when it comes to Trump's speeches, discussion of immigration, and physical attacks on Trump supporters.

One of the most hilarious and ironic posts I ever came across was when Zuckerberg gave a speech at Georgetown University about freedom of speech, and then Facebook content moderators were warned to watch out for and delete hate speech coming as a result of this speech:

Saadi Martinez shared a profile
6 hours [10/17/2019]

[Heads up][Mark Zuckerberg Live Speech]

Today, Thursday October 17th at 10am PDT/1pm EST, Mark Zuckerberg will be delivering a speech at Georgetown University on freedom of expression, that will be live-streamed from his personal FB profile (https://www.face-book.com/zuck). He will underscore the company's commitment to giving people a voice, while recognizing that we have a responsibility to remove content that has the potential to threaten safety or silence others. He will also touch on important movements that were initiated on Facebook such as Me Too and Black Lives Matter, and provide some examples of hate speech that are not permitted on our platform. Risk & Response and NA Markets have been partnering on risk mitigation steps in partnership with Product teams, but due to the nature of this commentary, we may see escalations or an increase in user reports of hate speech and wanted to provide a heads up on this. Please reach out with any indication of an increase in user reports or in volume.

[like] You and 131 others[1]

Very frequently, we were given similar kinds of warnings about possible violations whenever Trump gave a speech. Apparently, Facebook thinks anytime Trump opens his mouth, hate speech will come spewing out.

For example, on August 28, 2018, we were warned about a racist dog whistle from Trump:

Shawn Browder shared a link to the group: Cognizant North America Team

August 28, 2018

[Market Insights Request] -Trump's Tweet Regarding South African Farmers

Type: [Hate Speech/Credible Threat/Harassment Jobs]

Job Examples: N/A

Overview: FB is looking to identify any spikes in violent or hateful speech related to Trump's recent South Africa tweet.

The WHY: Trump tweeted about land reform in South Africa. Whether he knows it or not, this is a very strong dog whistle to American white supremacists, who associate the current issues in South Africa with the appropriation of land from white farmers in what was then Rhodesia and now is Zimbabwe. Bottom line is that this could easily become a very difficult fire with white supremacist content essentially endorsing a Trump tweet but making the innuendo more direct.

Article for context:

https://www.nytimes.com/2018/04/10/magazine/rhodesia-zimbabwe-white-supremacists.html

Request to Reps: Please surface edge-case/difficult jobs regarding Trump's recent tweet in the CONA Policy Tribe through SMES and TLs.[2]

There were many examples of whenever Trump said anything, they'd make an announcement to crack down on pro-Trump or anti-immigrant hate speech. This shows a selection bias. Are they doing the same whenever a Democrat makes a speech? Why is hate speech defined in such a way to inhibit right-wing talking points and speech?

Here is one such example of Facebook giving us instructions on how to treat political speech from Trump. This post was from roughly January 2019, and I filmed it on June 6, 2019:

Guidance given to content moderators:

Raquel Salinas* **Recap on Trump's Immigration Address**

SUMMARY
Donald Trump gave an address regarding the "humanitarian crisis" on the border. In his speech calling border security a "crisis of the heart," Trump argues status quo immigration policy is ineffective, negatively impacting minority communities, the opium crisis, overall American safety, human trafficking. Democrats responded by arguing that President Trump is holding the government "hostage" and should work to find a compromise.

POTENTIAL POLICY VIOLATIONS
As mentioned on this post, there may be user reports that come in after this address that may require consultation with the following policies:
Hate Speech policies (IS: 1.2 and particularly KQ 12 which was updated last October to give Central Americans the same protection as protected classes given their frequency with which they are targets of hate speech)
Credible Violence Policies(IS 1)

---

* A pseudonym.

ACTION ITEMS

An internal FYI is that anything stated by Trump himself, even if violating, would likely be allowed because of newsworthiness, but if there are questions on this please escalate so we can have Content Policy on-call weigh in.[3]

Here we see Central Americans were given protections as a protected class, even though they're from a region and wouldn't normally be protected (North Americans do NOT have protection). Specific nationalities are protected, but not all regions are protected unless specifically mentioned. Facebook states that they (Central Americans) are frequently targets of hate speech. Since Facebook was hyperfocused on immigration, it makes sense that they would notice hate speech toward Central Americans more than any other group.

And by conditioning its contractors to look for hate speech toward Central Americans, you're going to find more examples of it. However, many examples of "hate speech" are also "political speech," so Facebook is essentially censoring political speech if anyone posts about the border, even if that post does not contain an attack on the people crossing the border.

Many Americans are upset about illegal border crossings, and they direct that anger toward the people crossing and the politicians who encourage this to happen.

Facebook's own policy with regard to human smuggling is somewhat odd in that it allows people to ask instructions on how to cross a border, while prohibiting traffickers from finding potential customers. Here is a section of the Human Exploitation policy:

2. We do not allow any of the above mentioned forms of exploitation at any stage. We therefore remove:

1. Recruitment of potential victims which is the soliciting or targeting of individuals for the purpose of exploitation through force, fraud, coercion, enticement, deception, blackmail, or other non-consensual acts.
2. Facilitation of human exploitation by coordinating, transporting, transferring, harboring, or brokering of victims prior or during the exploitation.
3. Exploitation of humans by promoting, depicting, or advocating for it.

3. We do not allow content that offers or assists in smuggling of human[†]

---

† This was transcribed from the screenshot of the film I took, but the screen was cut off because of the camera angle when I filmed. Hence why it cuts here.

The type of content that does not violate our policies and is therefore allowed on our platform.

1. We allow requests for information or solicitation for help on how to get smuggled. [4]

Within the same policy, under Operational Guidelines we see the following:

D. What type of human smuggling content gets removed?

1. Any offer to smuggle or assist in smuggling a person, e.g.,

   1. specifying assistance
   2. "We'll help you get across"
   3. specifying a price
   4. "It will cost 700 per person"
   5. making guarantees of journey paths
   6. "safest routes"
   7. "avoid police"
   8. "I've paid off a guy"

E. What type of human smuggling content is ignored?

1. Provision of information on how to leave a country illegally, e.g.,

   1. Explaining how it's done:

      1. "If you go down to X beach you will able to find a boat"
      2. "Boats leave X beach on a daily basis with migrants"

   2. Explaining the gaps in Border Patrols:

      1. "The Irish Navy is set to remove patrols from July to October and that allows people to pass by undetected"
      2. "Because of red tape and bureaucracy, neither . . ."
      3. "Shift changes at midnight are often exploited"

   3. Criticizing the enforcement:

1. "Everyone knows you should wait until midnight"
2. "The coast guards in Italy are easiest to bribe"[5]

I find it shocking that Facebook allows people to discuss gaps in the Border Patrol. That seems like it violates the essence of the policy, which is to prevent criminal activity.

While discussing immigration, hateful phrases and racism are often evident in Facebook posts. So how should Facebook treat this content? It's clearly political speech, since it is discussing immigration, but at what point does it stop being political speech? Or does it ever stop being political speech?

Facebook gave us instructions on how to treat Trump's statement regarding immigrant gang members:

[PRFireRisk] #no
[Workflow] #CM
[Market] #NorthAmerica
[Type] #Hate
[SubType] #Dehumanizing

[Summary] On Wednesday 16-May, President Trump made a comment during a California Sanctuary Cities round table that immigrant gang members "aren't people" but "animals."

[Action] #multiple

[Question] Although the comments made by President Trump would be ignored due to referring to a subset based on violent crimes, there is the potential for hate speech directed toward QPCs on the platform.[6]

This quote by Trump was taken out of context, and many media outlets reported that he was calling immigrants at large "animals." However, it's clear he was referring to members of the MS-13 gang.[7] Facebook's policy does permit calling gang members "animals" but wanted to make sure we were protecting immigrants from hate speech.

Factually, Central Americans are breaking the law by crossing the border. And so Americans are using a public forum (Facebook) to attack this group of people for breaking the law. And then Facebook is stepping in as an arbiter of this political discussion and telling Americans that you can't attack groups of people for breaking the law.

When undocumented individuals cross the border, statistics show they are more prone to commit crimes. According to the Texas Department of Public Safety, between 2011 and 2020, hundreds of thousands of illegal aliens committed a wide variety of crimes:

> Between June 1, 2011 and March 31, 2021, these 231,000 illegal aliens were charged with more than 378,000 criminal offenses which included arrests for 698 homicide charges; 43,555 assault charges; 7,068 burglary charges; 46,293 drug charges; 606 kidnapping charges; 18,824 theft charges; 29,344 obstructing police charges; 2,067 robbery charges; 4,612 sexual assault charges; 5,743 sexual offense charges; and 3,955 weapon charges. DPS criminal history records reflect those criminal charges have thus far resulted in over 145,000 convictions including 309 homicide convictions; 17,194 assault convictions; 3,758 burglary convictions; 20,958 drug convictions; 228 kidnapping convictions; 8,043 theft convictions; 13,123 obstructing police convictions; 1,234 robbery convictions; 2,184 sexual assault convictions; 2,776 sexual offense convictions; and 1,542 weapon convictions.[8]

Additionally, "non-citizens accounted for 24 percent of all federal drug arrests, 25 percent of all federal property arrests, and 28 percent of all federal fraud arrests."[9]

The whole reason Facebook created a hate speech policy was to prevent real-world harm. Yes, racist attacks on immigrants can lead to physical violence. However, illegal border crossings can also lead to real-world harm. It is not racist to note that people with fewer financial resources are likely to be more desperate. Then there's the fact that drug cartels essentially control the border and who goes through. Mexico is not in control of its own border. Isn't it simple logic that such cartels would also provide safe passage for many of their members into the United States?

Therefore, how does Facebook determine which poses a greater threat?

Facebook specifically references Mill's Harm Principle as a basis for their hate speech policy. John Stuart Mill proposed a philosophy to preserve individual liberty while at the same time preventing harm to others:

> Mill distinguishes between "other-regarding actions" and "self-regarding actions." When other-regarding actions are such that they cause harm to others, it is appropriate to impose sanctions, in line with the harm principle.

However, this must be due to the harm they cause and not merely because they are found to be offensive or disgusting (i.e., laws should not be moralistic). With regard to self-regarding actions, laws should not be paternalistic.

Mill does place some limits on free expression of opinion, namely, when "the circumstances in which they are expressed are such as to constitute their expression a positive instigation to some mischievous act." As such, any expression which causes harm should be censored/outlawed. However, if the majority merely find a certain opinion disagreeable or disgusting this is something they must bear for the sake of the "greater good of human freedom."[10]

Using this rationale, Facebook should be actively looking for posts advocating for open borders and delete those posts, since open borders can lead to real-world harm, right? This just shows how much of a slippery slope Facebook's hate speech policy has become, and how subjective it is. Under the guise of protecting people from hateful words, almost any kind of political expression can be censored and suppressed. This is a dangerous precedent for free speech and can be used against liberals just as easily as it's currently being used against conservatives.

With regard to immigration, Facebook is taking sides by crafting its hate speech policy to suppress conservative talking points and political discussion. Here is some guidance given in May 2019 to content moderators:

[PRFireRisk] #no

[Workflow] #CM

[Market] #NorthAmerica

[Type] #Hate

[Subtype] #Dehumanizing

[Summary] On Wednesday 16-May, President Trump made a comment during a California Sanctuary Cities round table that immigrant gang members "aren't people" but "animals."

[Action] #multiple

[Question] Although the comments made by President Trump would be ignored to due to referring to a subset based on violent crimes, there is the potential for hate speech directed towards QPCs on the platform.[11]

Here Facebook is defending violent criminal gang members and arguing that these gang members should not be classified as "animals," but rather as "people." Per Facebook's policy, the words Trump used are "dehumanizing" toward gang members. QPC refers to a quasi-protected characteristic, someone who is protected some of the time under Facebook's policy, such as immigrants. Facebook went into great detail to determine which groups are considered criminal subsets and who are nonprotected subsets within the hate speech policy. Here is an explanation from the Operational Guidelines of the Hate Speech policy:

Other Criminal Subsets:

- Tier 2 except general terms for lawbreakers such as "criminals" including variations under Dehumanizing Speech, Other Criminals
- PC/QPC + Described as having carried out crimes (except violent crimes or sexual offenses)
- Ex. Illegal immigrant
- Ex. White Thieves

Non-Protected Subsets:

- No protection
- PC/QPC + Describes as having carried out violent crimes or sexual offenses
- Ex. Irish terrorists

Non-Protected Characteristics (NPC) (Not targeting a PC/QPC):

- No Protection
- NPC
- NPC _ Age
- Age
- Ex. Blondes, Rich children, children

C03: What is an "other criminal subset"?

An other criminal subset is a subset of a PC/QPC described as having carried out crimes (except violent crimes or sexual offenses). This category includes but is not limited to "illegals," "illegal immigrants," "illegal aliens," "undocumented migrants," "unauthorized migrants," "clandestines," and "PC + thieves/crooks/etc." (e.g., "American crooks"). Other criminal subsets are protected from Tier 1 attacks except referring to the subsets as "criminals," "offenders," "felons," or other general terms for law breakers. They are protected against comparisons to specific types criminals (e.g., "thieves," "drug dealers," "murders," "rapists," or "terrorists"), however.

No action:

- Illegal aliens are criminals
- Illegal immigrants are felons
- Illegals are crooks
- German thieves are criminals

Delete:

- Illegals are thieves
- Migrants are criminals
- Illegal immigrants are terrorists
- Immigrants are crooks
- White criminals are filth[12]

So we see that Facebook modified their hate speech policy to allow some political speech discussing immigration beginning in 2019. But you still can't call illegal immigrants thieves or call migrants criminals. I understand it's difficult to regulate speech, but the average person discussing or parroting Republican talking points won't understand this complex nuanced system of policing speech, and the end result of Facebook's policy is the erasure of political discourse.

Here is another piece of guidance dating all the way back to November 2017:

> Hardeep Jandu shared a link to the group: Cognizant North America High-Pri Updates
> November 29, 2017
>
> #HeadsUp
>
> Donald Trump retweets anti-Muslim videos posted by Britain First deputy leader

Hi team, Sending over a quick note as a heads up regarding current events. Please be extra aware when reviewing for potential hate speech and credible violence violations.

https://news.sky.com/story/donald-trump-retweets-anti-muslim-videos-posted-by-britain-first-deputy-leader-11148844"[13]

This is an example of Facebook prioritizing Trump for potential hate speech violations. The video he shared showed someone committing property damage. But Facebook told us to watch out for hate speech, not to look for encouraging property damage, which is covered under the Coordinating Harm policy.

We were also told to prioritize content related to the migrant caravan from Honduras as well as content related to George Soros:

Jessica Wyn‡ shared a link to the group: Mid Term Elections-Voter Information & Voter Fraud Jobs Submission
October 25, 2018

[High Priority] [Job Examples Request]-George Soros, Caravan, Pipe Bombings

Hello everyone!
The client has requested for IDs on **ANY** content related to **George Soros**, the **Caravan**, or the **recent pipe bombings**. Please post them in the comments with descriptions and your action on it. Thanks!

- Pipebombs/Explosive Devices updated guidance in comments (click for guidance)
- Cesar guidance (click for guidance)

#GeorgeSoros #Caravan #Pipe #Bombs #Terrorism #Cesar[14]

Here is yet another example telling us to look for examples of hate speech against the caravan from Honduras.

"HONDURAS SEEKING TO REACH THE US BORDER
[Category] HeadsUp

‡ A pseudonym.

[PrFireRisk] Yes
[Workflow] Multiple
[Market] ESLA

[Summary] Reports indicate another caravan of Central America migrants is forming in Honduras seeking to reach the U.S. border. Thousands are expected to reach Mexican soil in the coming days. President Donald Trump announced the US would be cutting funding for El Salvador, Guatemala, and Honduras, known as the "Northern Triangle," in response to the Trump administration's perceived lack of effort to mitigate the "migrant crisis."

[Action] There is an increased likelihood of an uptick in Hate Speech on the platform against Central American migrants (PC). Please be on the lookout for potentially related violations around our Coordinating harm, Credible Violence, and Dangerous Orgs policies.

Please share examples and trends related to this event

Thank you!

http://time.com/5561225/mexico-central-american-migrant-caravan"[15]

Once again, Facebook is giving guidance to its moderators to "be on the lookout" for hate speech related to the caravan. But the hate speech didn't just apply to Trump mentioning immigrants; it also applied to his recognizing Jerusalem as the capital of Israel:

Guidance regarding Jerusalem

Shawn Browder shared a link to the group: Cognizant: North America High-Pri Updates

December 5, 2017

#Newsworthy #PossibleTrending #HeadsUp

TRUMP MAY ANNOUNCE JERUSALEM AS ISRAEL'S NEW CAPITAL

Summary: Reports indicate President Trump may announce
the United States' recognition of Jerusalem as Israel's capital this week.
The move is largely a symbolic gesture intended to fulfill a
campaign pledge. Trump's impending decision on Jerusalem touches on
one of the most sensitive issues of the Israeli-Palestinian conflict.
In response, Hamas (the Palestinian Islamist political organization)
warns of potential retaliation, increasing likelihood of protest
activity and violence in response to the announcement. The decision
is expected to cause anti-Israel and anti-US sentiment in Muslim
majority countries, including on social media.

Please note Zionists are considered a vulnerable group. Any call for action
of violence, statements advocating violence against, or aspirational/conditional
statements of violence targeting Zionists should be
deleted.

Potential Policy Violations: #HateSpeech, #CredibleViolence"[16]

It's fascinating to see Facebook's own prejudice and political bias seep into these posts. "The move is largely a symbolic gesture intended to fulfill a campaign pledge." That phrase reeks of partisanship and is a slap in the face to the millions of Christians and Jews who consider Jerusalem a sacred city and part of Israel.

Both Presidents Clinton and Obama recognized Jerusalem as the capital of Israel, and this action by Trump is simply following through on the historical reality of the city.[17]

However, this post also did point out the possible backlash against Zionists, and Facebook has extra protections for Zionists as a "vulnerable group," which is admirable and commendable.

* * *

As you can see, the migrant caravan topic came up a lot. It was also in the news nonstop during that time period in 2018. This was a huge theme for Trump because it dealt with immigration and the fact that he campaigned on building a wall.

Every time Trump mentioned immigration, Facebook immediately told its moderators to watch out for hate speech violations. When Trump referred to gang members as "animals," Facebook was quick to point out that "immigrants" have special protections under their policy. In general, it is very difficult to discuss immigration on Facebook.

Under Tier 3 of the Hate Speech policy, no exclusion is allowed. So if I say, "Keep Canadians out of the United States," that violates Facebook's rules.

Referring to any nationality as uncivilized or barbaric is against the rules at Facebook. If I refer to the Goths or the Vikings as "barbaric," this is a violation of the hate speech policy. The 2004 animated television series *Dave the Barbarian* would not pass the smell test and would be considered for review by content moderators such as myself. Such characterizations of groups of people are considered "dehumanizing."

Tier 1 of Facebook's Hate Speech Policy

The type of content that violates our policies and is therefore NOT allowed on our platforms:

1. Tier 1

Content targeting a person or group of people on the basis of their protected characteristic(s) or immigration status (including all subsets except those described as having carried out violent crimes or sexual offenses with:

1. Violent speech or support for death/disease/harm in written or visual form
2. Dehumanizing speech or imagery in the form of comparisons, generalizations, or unqualified behavioral statements to or about:

    1. Insects (including but not limited to: cockroaches, locusts)
    2. Animals in general or specific types of animals that are culturally perceived as intellectually or physically inferior (including but not limited to: vermin, pig)
    3. Filth (including but not limited to: dirt, grime), bacteria (including but not limited to microbes, viruses), disease (including but not limited to: cancer, sexually transmitted infections), feces (including but not limited to: shit, crap)
    4. Subhumanity (including but not limited to: Savages, Devils, Monsters, Primitives)
    5. Sexual predators (including but not limited to: rapists, child molesters) . . .[18]

Hate Speech Policy—Tier 3 Exclusion

"Known Questions" is a supplement to the policy and is somewhat of an internal FAQ for content moderators.

KQ RR: How do we define exclusion categories?

- Calls for exclusion targeted at a protected category, these include explicit exclusion and calls to limit political, economic, and social rights to a protected category as defined below

  ○ Explicit exclusion including but not limited to:

    - Kick out/Expel + PC
    - No/No more/Not allowed/World without + PC
    - Don't have rights/Shouldn't have rights + PC
    - Advocating for racial purity

  ○ Political Exclusion: denial of right to political participation, including but not limited to:

    - Right to organize (e.g., Women should not be allowed to protest)
    - Right to Vote/ Suffrage (e.g., Black people should not be allowed to vote)
    - Candidate eligibility (e.g., Gay people should not be allowed to run for president; Do not vote for this candidate because she is black). Please note: does not include statements about personal voting preferences such as "I couldn't vote for a black candidate" or "I will never vote for a Muslim candidate," which are allowed.

  ○ Economic Exclusion: denial of access to economic entitlements and limiting participation in the labor market, including but not limited to . . . [19]

It's interesting to see that if I'm talking about illegal aliens who cross the border, I can't even say that they shouldn't have rights in the United States. They don't have rights in the United States because they're not citizens. But I can't express that while on the Facebook platform.

Other Notable Aspects of the Hate Speech Policy

A Protected Characteristic (PC) is defined as someone's race, religion, ethnicity, gender, sexual orientation, national origin, disability, etc.

If you are attacking someone based on their PC, the applicable policy is the "Hate Speech" section of the Implementation Standards (IS). The abbreviation often used for the Hate Speech section is HS. It is divided up into tiers. If you are attacking Canadians and dehumanizing them by calling them terrorists, animals, subhuman, or sexual predators, that falls under Tier 1 (the highest and most severe tier).

Tier 2 of the Hate Speech policy covers calling Canadians dumb, retarded, idiotic, etc. Tier 3 of the Hate Speech policy refers to excluding Canadians from any process or physically excluding them from any civic activity.

In late 2019, Facebook modified their hate speech policy and made it more nuanced to allow more statements about specific events.

This is called a "qualified behavioral statement" as opposed to an "unqualified behavioral statement." For example, prior to 2019, if you said "Muslims did 9/11," this would be deleted.

After the changes, this was allowed because it is referencing a specific historical event.

Screenshot of a Clarification for the Hate Speech Policy

"Antonio LaSalle[§]
July 10 2019

#hate #weeklyquestions #processclarifications

1. How are we to action attacks that target a numerical minority of a PC (49% of white people)?

   ○ Numerical minorities of PCs do not get protection under the hate speech policy outside of designated dehumanizing comparison. Numerical minority can be established through terms like some, few, or any number that would be less than 50% of a PC.

   2. How do we interpret content that is calling illegal immigrants criminals?

---

§ A pseudonym.

○ If content is generalizing that a majority of immigrants are criminals, we delete for dehumanizing. However, stating that illegal immigrants broke the law by entering a country illegally or calling immigrants criminals because they are illegally entering a country would be ignored, as it is only reporting on an action and not an overarching generalization.

#915116158840099 – Ignore
#2262907120431368 – Ignore
#639988236484251 – Ignore

[reactions] You and 483 others   seen by 571"[20]

Trump's State of the Union Address

On January 30, 2018, Trump delivered his second State of the Union Address.

On January 31, 2018, Deeprek Hundai⁵ posted in the Cognizant: North America High-Pri Updates the following information:

#HighPri #HeadsUp #USElections

State of the Union Address—Trends/Topics

Hi everyone. Sending over a note on the US State of Union address. If you see emerging trends, please work with your TL to raise them up to conapolicy. Thanks!

POSSIBLE TRENDS WE EXPECT TO SEE IN QUEUES:

- Hate related to illegal immigrants (Dreamers, DACA), immigrants in general
- Hate related to Israelis, Jews, Trump's recognition of Jerusalem as Israel's capital.
- Hate related to North Korea regime but possibly its citizens
- Cruel mocking of death/disability of Trump's guests (see below)
- Hate or PSR related to prisoners of Guantanamo Bay

---

⁵ A pseudonym.

PHRASES OR PEOPLE YOU MIGHT SEE:

- **"Down the middle"** -bipartisan plans for immigration reform
- **"Chain migration"** -process where a migrant can bring over relatives
- Considered offensive by migrant communities, "family reunification" used instead
- **"America First"** -Trump's motto for prioritizing domestic needs before those of other countries, including allies
- **"Visa lottery"** -offers visas to immigrants from underrepresented countries; many are the nations Trump referred to as "shithole countries"
- **Guantanamo Bay** -military prison for terrorism-related offenses deemed as breach of human rights by Amnesty International
- **MS-13** (FB designated DO[dangerous organization]) -Gang ethnically composed of mostly Central Americans

   ○ Trump highlighted MS-13 violence by illegal immigrants against young victims Kayla Cuevas and Nisa Mickens (public figures)
   ○ Victims' parents: Evelyn Rodriguez, Freddy Cuevas, Elizabeth Alvarado, and Robert Mickens

- **Otto Warmbier** (public figure) – American student who died after being sentenced to imprisonment and hard labor in North Korea

   ○ Parents and siblings present at SOTU, Trump described as a "hardworking student"

- **Ji Seong-Ho** (public figure) – North Korean defector who lost limbs while escaping regime, now lives in South Korea and is an activist

   ○ Praised by Trump as hero, seen in SOTU holding up old crutches and standing on prosthetic leg

IMPORTANT THEMES:

- **Immigration reform in 4 pillars**, praising ICE and border patrol for service

   1) Making path to citizenship for illegal immigrants who meet education/work requirements and show good moral character over a 12-year period

2) Building southern border wall, ending "catch and release"

3) Ending visa lottery, moving toward a merit-based system

4) Ending "chain migration" by limiting to only immediate families. Democrats hissed and booed this pillar.

- **Gang violence, particularly MS-13**

  ○ Examples given of tragic deaths from gang violence by illegal immigrants

  ○ Trump called on Congress to fix "glaring loopholes" to enter the country as unaccompanied alien minors.

- **Protecting American workers against unfair trade deals**

  ○ Era of economic surrender is over. Trump vowed to protect American workers and intellectual property through strong enforcement of trade rules.

- **Keeping open Guantanamo Bay**

[REACTION: LIKE, HEART SURPRISE]   Seen by 512[21]

Here we see guidance toward Trump's State of the Union speech. The guidance tells us to "raise up" emerging trends to our Team Leader (TL), so that the TL can raise it up to CONA policy. This guidance has some unbiased aspects to it. For example, it says to watch for "cruel mocking of death/disability of Trump's guests," even if one of those guests was a defector from North Korea. Also, it tells us to look for hate directed toward Israelis and Jews so that we can prevent those kinds of verbal attacks/hate speech against them.

It also mentions that we should look for hate or PSR related to prisoners of Guantanamo Bay. The PSR (Praise, Support, Represent) I understand. This is a term used in conjunction with the Dangerous Orgs policy. This is telling us to watch for people praising terrorists who are being held at Guantanamo Bay. However, it's also asking us to look for hate directed toward terrorists. I'm not sure why Facebook is worried about protecting terrorists and looking for hate speech directed toward terrorists. Shouldn't murderers be hated? Or is Facebook trying to change that natural human inclination, as well?

We also see a list of phrases including "America First," "chain migration," etc. Some of the small adjectives and words hint toward bias in the definitions of these phrases. In the definition for "visa lottery," the author of this post goes out of their way to include a reference to Trump's use of the phrase "shithole countries." Also, the definition of "chain migration" says that this phrase is offensive to migrant communities. And last, for the phrase "America First," they define it as prioritizing domestic needs above those of other countries, including allies. The way this is worded implies that America is shirking its responsibilities toward its allies.

In this post, I do see some warnings about hate speech, but these are all Trump's political talking points. Once again, this reaffirms the fact that Facebook does censor political speech, contrary to what Mark Zuckerberg testified to in April 2018:

> **CRUZ:** Let me try this, because the time is constrained. It's just a simple question. The predicate for Section 230 immunity under the CDA is that you're a neutral public forum. Do you consider yourself a neutral public forum, or are you engaged in political speech, which is your right under the First Amendment?

> **ZUCKERBERG:** Well, senator, our goal is certainly not to engage in political speech. I am not that familiar with the specific legal language of the—the law that you—that you speak to. So I would need to follow up with you on that. I'm just trying to lay out how broadly I think about this.[22]

Facebook went to great lengths to determine what kind of speech was considered "political speech" and would thus be allowed. For example, if you call someone a "Trumphumper," this was considered an "exaggeration of speech" and is allowed.

But calling someone a "snowflake" is deleted.

This also applies to comparing Trump supporters to the KKK. Here is guidance given by Zachary Ferrier, a Cognizant employee and member of the Phoenix policy team:

Zachary Ferrier to Cognizant North America

August 14 [2019] at 3:28pm

#Trending #KKK #Pink #MAGA #Laundry

[ID] 596008744139310

[Description] An image of three KKK members in uniform, but one is pink instead of white with a caption making a joke about mixing their robes and MAGA hats in the laundry.

[Action] Ignore

[Reasoning] This is expressing a political comparison by implying that Trump supporters are also part of the KKK.[23]

So you can compare Trump supporters to the Klu Klux Klan, but what does the Bullying Policy say about attacks on individuals?

What is considered to be a negative character claim?

A. Specific terms or definitions that attack an individual's mental or moral qualities. This encompasses: disposition, temperament, personality, mentality, etc. Claims solely about an individual's actions are not encompassed, nor are criminal allegations.

John Smith is the least intelligent person I know = remove with name match

John Smith is an idiot = remove with name match

John Smith failed math = ignore, even with name match

John Smith failed math = ignore, even with name match.

John Smith is really fucked up = remove with name match.

John Smith is a robber that everyone should be aware of = ignore, even with name match

Note: Claims about ideology or behavior such as "racist," "xenophobe," "homophobe," "fascist," "nazi," "misogynist" should not be considered negative character claims.[24]

That last part where it allows calling people racist Nazis was added in 2019. I don't know about you, but if someone calls me a racist, that's a personal affront. Oddly enough, if someone calls me "kind and considerate," and I report it directly, that would also be deleted.

But they can call me a Nazi racist all day, and it will never get taken down.

Returning to the KKK example, if someone photoshopped my picture with a KKK robe and hood and called me a Trump supporter, that would be allowed. Despite its attacking my disposition, temperament, and mentality, Facebook carved out an exception to call people racists. Also, Facebook's policy regarding hate groups such as the KKK is very strict:

Dangerous Individuals and Organizations policy

1. We do not allow the follow people (living or deceased) or groups to maintain a presence (e.g., have an account, Page, Group) on our platform:

    1. Terrorist Organizations and terrorists
    2. Hate organizations and their leaders or prominent members
    3. People notable for physically attacking people based on a protected characteristic
    4. People who have committed or attempted mass murder
    5. People who have committed multiple murder
    6. Criminal organizations and their leaders and prominent members

2. We do not allow symbols that represent any of the above organizations or individuals to be shared on our platform without context that condemns or neutrally discusses the content.

3. We do not allow content that praises any of the above organizations or individuals or any acts committed by the above organizations or individuals.

4. We do not allow coordination of support for any of the above organizations or individuals or any acts committed by the above organizations or individuals.[25]

As you can see, hate organizations like the KKK cannot have a presence on the platform. There can't be any coordination of support for these organizations either, as we see under point number 4.

The only reason this political meme about the KKK is allowed is that it is sharing condemnation or neutral speech regarding the organization. This is only true if you assume that Trump supporters are bad, though, and that the meme is an attack on Trump supporters. It's like saying, "Trump supporters are just as bad as the KKK."

If the meme was interpreted as a positive reaffirmation of Trump supporters and the KKK, it would be deleted. By allowing this meme, you have to operate under a lot of assumptions. Otherwise, you violate the Dangerous Individuals and Organizations policy by lending support to the KKK.

Cartoon Attacks of Political Figures

Another modification to Facebook's policy prohibited cartoons that showed political figures being attacked.

- [Cartoon image of Trump kicking Obama down a well]
  "Violates since it depicts a fatal method of killing a living human target (kicking down a deep well)."[26]

This is important because it's a recent change (the change went into effect October 2, 2019) in how we action visual threats against public figures. Previously, we ignored all visual threats against high-risk individuals. However, now certain cartoon imagery is a delete. This can be used in the election against conservatives who are depicting political speech against candidates, and their content can be more easily deleted. It can also be used against liberals; however, the fact that Facebook can change their rules anytime they want means they can adjust their policy depending on the election year and who is running for office.

Offering Money to Attack Trump Supporters

On June 6, 2019, I filmed guidance from Facebook about a trending post. The post was from August 20, 2018, and stated the following:

Mariana Sanchez to Cognizant ESLA Trending
August 20, 2018

#Trending #Ignore #MAGACap

232857230754115

Video of a man offering his daughter $100 to knock of MAGA cap off a boy's head

Action: Ignore

[reaction] You, Jessica Martinez,** Patricia Charcon,†† and 99 others[27]

This type of violence (knocking/hitting) is considered midseverity violence per Facebook's Violence & Incitement policy (V & I). I have no explanation as to why Facebook would tell us this does not violate the policy, especially if it were shared with a praising or encouraging caption.

Viral Video Deleted Showing an Attack on a Trump Supporter

Additional guidance was given on July 6, 2018, regarding a viral video showing a Trump supporter being attacked. I've already mentioned this incident, but here's the specific guidance we were given on the video:

Mariana Sanchez to Cognizant CO Trending
July 6, 2018

#Trending #Delete #Bullying #TargetedCursing

41713666547241

Man steals MAGA hat from a 16-year-old boy and also throws a drink at him. He then proceeds to curse the boy out. Since we have context that the cursing is directed toward the 16-year-old, we will **delete** this content for bullying a minor.

[reaction] You, Jessica Martinez, Patricia Charcon,‡‡ and 445 others[28]

This was a viral news story and was reported on by Fox News[29] and many other media outlets. In many instances, the curse words were bleeped out

---

** A pseudonym.
†† A pseudonym.
‡‡ A pseudonym.

of the video. Yet Facebook made a decision to delete this video across the board.

This video would have definitely drawn sympathy for Trump supporters and shows the vitriol and hatred that many Trump supporters receive for simply showing their support for President Trump.

The video shows a man forcefully removing a MAGA hat from a teenager. Since the victim was a minor, and the policy doesn't allow cursing at minors, guidance was given to delete this video. However, any condemnation context with a bullying violation should be allowed, and this caveat wasn't specified in the post. All the content moderators aligned to delete this content. The reason they post things is so we're all actioning the same way. Here is some of my personal analysis of this situation:

- Cursing at a minor is a delete without a name or face match. Primarily we see this in text-based posts where we have a target, ex. Curtis, and they're calling him a dick or an idiot.
- However, the spirit of the policy is to prevent harm, according to Mill's Harm Principle. Resharing a video of an attack doesn't in and of itself violate Mill's Harm Principle. But if you're sharing it with the intent of mocking the individual, or the share caption is supporting the violation, then it should be taken down.
- If there is a condemning caption, then a video share should be allowed. But when people watch the video the main point isn't the cursing at the minor. The key point of the video, is the attack on the MAGA-hat victim. The cursing at the victim is ancillary.
- When you watch the video, the cursing isn't directed at you, and deleting an example of bullying doesn't help to raise awareness against bullying that occurs, regardless of your political bias.

I'm not entirely sure there was even a violation of the Bullying Policy at this point. There were some changes to the Bullying Policy throughout the years. But I don't see a clear violation, especially if the video was shared with a condemning caption.

For example, if I shared this video with a link to Fox News and added a caption saying, "how horrible," then I'm condemning the attack, and my post should be allowed on Facebook. Facebook says they're deleting it because the cursing is directed toward the minor, but as a Facebook user, I am not cursing at the minor, I am simply sharing a video of an instance of cursing at a minor.

Facebook's rationale is very weak, because it implies that any videos that contain any curse words toward a minor should be deleted.

Section 6 of the Bullying Policy states:

[Not Allowed] 6. Target private individuals who are minors with:

    1. Allegations about criminal or illegal behavior

    2. Videos of physical bullying or violence in a fight context shared in a
       non-condemning context[30]

So according to the policy, I cannot share videos of physical bullying or violence with a noncondemning context. First of all, this part of the policy only applies to fights between minors. Second of all, I then should be able to share this video as long as I condemn the bullying/violence from the video. Is this the real reason Facebook banned this video? Because it would have had to be accompanied by some sort of caption that read, "Isn't it terrible how these Democrats are bullying a young Trump supporter?" I don't think that was the kind of narrative Facebook wanted, so with their unlimited power, they were able to simply delete the incident from the public conversation.

    Last, deleting this video doesn't fit the spirit of the policy. In another decision, we received guidance from Facebook that:

> The client has given us guidance of "Though an argument can be made for a
> letter of the policy delete, the spirit of the policy here would not be a delete.
> Please ignore."[31]

Essentially, the spirit of the policy for bullying is to prevent harassment of individuals. How does deleting this video prevent harassment? Who is being protected by deleting this viral video? The only purpose served by deleting this video is to prevent sympathy toward a Trump supporter who was brutally attacked and squelch a broader conversation about whether the liberal left was resorting to violent tactics against its political opponents.

Summary

In closing, my perception was that Facebook treated any instance of Trump opening his mouth as hate speech. This treatment also trickled down to Trump's supporters. Whether it's disallowing any discussion of immigration under the Hate Speech policy, or deleting viral videos of Trump supporters getting attacked, it's clear that Trump and his supporters were not given a fair shake.

Although Facebook made some slight changes in 2019 to allow for more types of speech (ex. "Muslims did 9/11" is now permissible), I believe these temporary changes were made to appease conservatives during the one-year civic audit[§§] performed by the Covington Law Firm and finalized in August of 2019.[32] Facebook can, at a whim, modify any of their policies, at any given time.

Facebook's true colors are highlighted by the fact that they advised content moderators to look for hate speech arising from historic occasions such as President Trump's 2018 State of the Union Address and his recognition of Jerusalem as the capital of Israel. (Facebook claims to be against bullying and for discussion of political issues, but it's clear that doesn't apply if you depart from their liberal orthodoxy.)

If they feel so comfortable canceling us, we shouldn't feel the slightest hesitation about canceling them.

[§§] The civic audit is examined in greater detail in Chapter 18.

Trumphumpers and Feminazis: Facebook's Slanted Approach to Bullying

Facebook's bullying policy is designed to prevent attacks on one's character, among other things. Within this policy we also find protections against the doxing of one's address and a wide variety of threats to online safety. However, it can be misused against conservative viewpoints, which is what I will cover in this chapter.

The Bullying Slang List

The Bullying Slang List was created as a resource by Facebook to help content moderators be aligned in how they action common words and attacks. I have several screenshots of the list, which includes more than a hundred words.

Below is a list of some of the words and attacks. The bullying policy is designed to prevent attacks on one's character, among other things, I believe the bullying policy reveals the greatest single weakness of Facebook's current strategy. From an extremely young age, human beings are wired to rebel against injustice. Put two toddlers together, give only one of them a cookie, and observe the mayhem that ensues. We are not just sensitive to injustice against ourselves, but we also stand up and complain when we see it done to others. This trait is universal across all cultures. We understand that societies function best when there are rules, but they need to be applied equally. If they are not, rebellion will inevitably follow.

In order for all of the content moderators to action content the same way and be in "alignment," we used Facebook's resource, the Bullying Slang List. This list was added to and updated periodically. On August 22, 2019, on-site policy manager Shawn Browder shared Version 3 of the list on SRT workplace as a PDF;

Shawn Browder uploaded a file in the group: Cognizant North America Team

August 22, 2019

"[Bullying Slang List v3] [Updated]

After revisions with Bullying & Harassment Focus SMEs across North America sites, FB is excited to launch an updated slang terms list that documents common slang terms and their associated decisions. These terms reflect the most current policy language and understandings. Please note this list is not exhaustive, and you are able to utilize Urban Dictionary if a term is not found on this list. If you have any questions, please feel free to reach out to your site's Bullying focus SMEs!

CC: LESLIE, ORION, LY, KATELYN, MATTHEW, YESENIA, SUZANNAH"[1]

Here are some of the words from the bullying slang list, which I filmed and documented:

Troll = negative character claim
Attention/Internet whore = negative character claim
Ignorant = negative character claim
Pedophile = ignore
Nazi = ignore
Trumphumper = ignore
Gender confused = claim about gender identity
Fucktard = negative character claim
Libtard = negative character claim
Trumptard = negative character claim
Racist/Sexist = ignore

Feminazi = negative character claim
Snowflake = negative character claim
Bigot = ignore[2]

For example, if I call John Smith a pedophile and he reports it, the comment stays up no matter what. This is a reported comment with a name or face match. I called John Smith a name, and he directly reported it. As a content moderator, I see the name of the reporter on the screen, along with the reported content. I can also see the "parent content," which just refers to the comment above, so I have context as to why he was calling him a pedophile.

Allegations of criminal activity are allowed unless I'm calling them a derogatory sexual term such as "whore" or "slut." Therefore, I can accuse someone of being a thief and that's allowed, even if they report it.

From the list, we see that there is some equality with regard to terms like Trumptard and Libtard. Both get deleted.

The bias starts to creep in with the term "snowflake."

This is such a benign term, and I've seen it used synonymously with the term "SJW."

At least for me, calling someone a snowflake is used to attack liberal ideology. However, if I call John Smith a "snowflake" and he reports it, the comment would be deleted. If in my next comment under John Smith's post, I call John a "Trumphumper," it would stay up forever *even if* John himself reports it.

Also, deleting the word "troll" is kind of weird. This seems like a very ubiquitous term that describes someone's trolling behavior on the Internet.

To summarize, attacks like "snowflake" and "feminazi" get deleted when reported. But if I call someone a "Nazi" or a "Trumphumper," there is no way to delete that.

As you can see, the only fairness we see is that calling someone a Trumptard is a delete.

All the other attacks on right-wing people such as racist/sexist/nazi/trumphumper are an ignore (allowed). Conversely, lefties have additional protections, because you can't call someone a snowflake or a feminazi (these attacks get deleted), in addition to calling someone a libtard (deleted).

The most interesting part is you can call someone a Nazi, but you can't call someone a Feminazi. This means Nazi feminists have additional protections.

Online Bullying of Children

Name-calling is nothing new to society, and anonymity seems to have bred and encouraged even more attacks on individuals. This is unfortunate, and I condemn it.

Bullying and harassment online is a problem for children and youth. Facebook has done a good job of creating this policy that protects children. Targeting of schoolmates has led to depression, decreased performance at school, and even suicide. Youth nowadays spend even more time on social media and use it as a primary means of communication with their peers. Even one's sense of self-worth is tied to the use of social media. Trolling others online has become a sport.

Research shows that online bullying has severe real-world consequences:

> Online hate, albeit conducted in the virtual world, may have dire real-life consequences at both individual and population levels. For example, the cyberbullying among youth and student populations and subsequent links with poor mental health, depression, trauma, substance misuse, and a higher risk of suicide are well-documented.[3]

Since social media is addictive, and so many minors use it, it's become a place where bullying and targeted attacks are prevalent.

Parents can help their children by limiting the amount of time they spend with their phone or on the computer. There are also many parenting apps that can control and monitor apps and cell phone usage.

Bullying is a major problem for children in elementary school, middle school, and high school. Therefore, what is Facebook's role and what are they doing to prevent bullying?

Within Facebook's bullying policy, more rules exist against bullying minors. For example, any attack on a minor's character is deleted, regardless of who reports it. For adults, this normally would require that the same person being attacked report the comment.

I also deleted many direct messages on Instagram that were attacks against minors. However, the minor still saw the hateful message before it was reported and taken down. Additional protections for minors is one of the few areas Facebook should continue to improve and is one of the main purposes for which Facebook was given legal protections to censor speech.

The Communications Decency Act was designed to protect minors on the Internet, so by protecting minors, Facebook is acting within the law.

However, what I did find surprising was the gray area that existed with regard to Facebook's enforcement of child pornography. To escalate a piece of content to Facebook that we suspected of being child porn, it had to meet certain criteria. Just an image of a minor in a skimpy bathing suit wasn't enough to escalate or even to delete.

They did train us to estimate the age of the minor. Obviously, if there was any clear nudity (genitalia), we were told to delete the image. However, there could be images of girls between the ages of six and ten wearing a skimpy swimsuit and posing in a provocative manner that were not deleted and did not fit the criteria.

While I was at Facebook, they told us we were overescalating, or sending too much suspected child pornography to the Facebook employees. I find this reprehensible that Facebook can prioritize hate speech over child pornography, because there was clearly a great amount of resources dedicated to hate speech. If there was an overflow of child pornography, couldn't Facebook reassign employees to that queue?

Which is more likely to cause significant harm to society?

I believe Facebook should have given content moderators more ability to act on our own judgment when it comes to child pornography. We know what "looks creepy" and should be able to act on that. But we were not allowed to.

The definition of "sexually suggestive pose" that Facebook gave us in the CEI (Child Exploitative Imagery) training deck was very specific. But we were not allowed to escalate something that we "thought" looked like child porn.

It had to meet the criteria.

For example, some Facebook Groups I reviewed had a profile pic of a sexually posed ten-year-old in a swimsuit. The name of the group was something innocuous and vague like "Great Pictures." For me to take down the group, the title, and the description of the page had to have some sexual element, or at least 30 percent of the randomized posts I saw had to violate the policy. Otherwise, it would not be taken down.

Sexual perverts and purveyors of this filth are notorious for skirting the rules and knowing how to evade Facebook's policies. By having strict rules for the content moderators to follow, the rule-breakers were able to adapt and know exactly how to hint at sexual content of minors without getting their content removed. If Facebook were to give content moderators more leeway, then more questionable pages could be removed.

The acronym CP, standing for Child Porn, is used in both North America and Latin America. If they shared that acronym and hinted at sharing images or asking to be contacted, we could delete that.

But as I've stated earlier, our hands were tied when it came to content that looked "off" or "creepy" but clearly didn't violate the CEI policy. I wish Facebook had given us more latitude to delete any kind of imagery that showed minors in any state of undress.

Just as with other sections of their policy, Facebook can also make exception to the child pornography policy and allow it in certain instances. For example, there is a famous photo from the Vietnam War showing a nude child running away from the bombing and destruction caused by the war. This is allowed. But one exception they gave surprised me:

Cindy Fredrickson* to Cognizant: North America Team
January 30, 2018
#Exception #ChildNudity
Newsworthy Exception

There is a trending image of an album cover for the Brazilian band Negritude Junior. The album cover does violate our Child Nudity policy. However, given the artistic value of this album cover as well as the public interest value, Facebook has decided to make a newsworthy exception.

The post itself as well as reshares by regular users should be ignored. However, CE labeling (Naked Babies) still applies, as the cover depicts fully nude toddlers.

If there are any doubts about how a particular piece of content is shared, please flag it to your SME or TL for review.[4]

The image shows roughly four or five children completely nude, and they're not babies. Their age is roughly four to five years old. I was shocked when I saw that Facebook could made exceptions to the child nudity policy for the sake of "artistic" and "public interest value."

Of all the policies that Facebook should address the most proactively, child pornography should be the highest priority. By employing human censors but not allowing them to use judgment, Facebook damaged children's lives and enabled the proliferation of child pornography on its platform.

---

* A pseudonym.

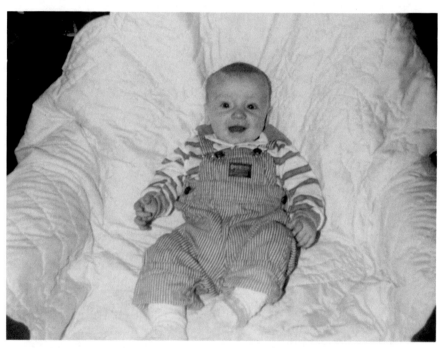

An early photo of me.

Ryan takes a break to play Jenga while an employee at Cognizant in Phoenix, Arizona. Photo taken in 2019.

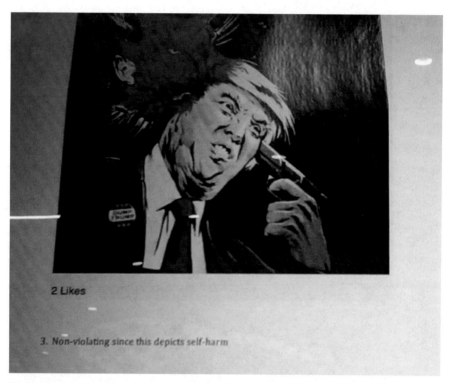

2 Likes

3. Non-violating since this depicts self-harm

This is considered cartoon imagery of a suicide, which at one point wasn't deleted under the suicide policy. Because he is shooting himself instead of someone else shooting him, it was allowed.

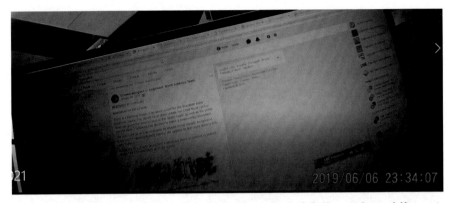

Facebook gave an exception to allow imagery of fully nude toddlers. A Brazilian children's music group featured this on their album cover, and Facebook decided that it had artistic and public-interest value.

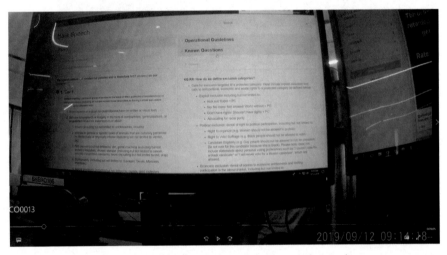

This image shows Facebook's Hate Speech policy. The exclusion categories prohibit users from saying, "Keep Canadians out of the US," or from saying, "Men shouldn't be making laws about women's bodies."

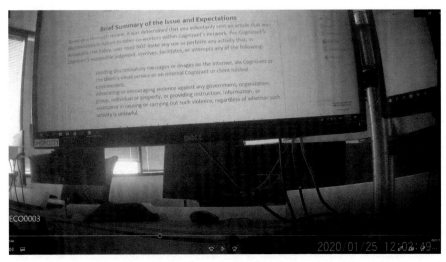

This image shows the PDF Corrective Action (CA) punishment I received from HR for sharing a link to an *LA Times* article, which featured me on the cover protesting the murder of the Charlie Hebdo satirists by radical Islamic extremists.

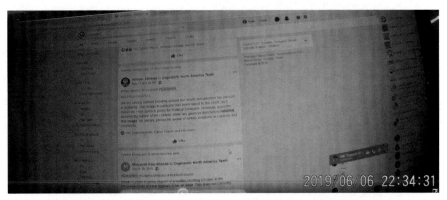

This is an image of guidance we received from Facebook regarding Alabama's abortion law. Pro-choice activists were violating FB policy by saying, "Men shouldn't be allowed to make laws about women's bodies," but FB gave an exception to allow that phrase.

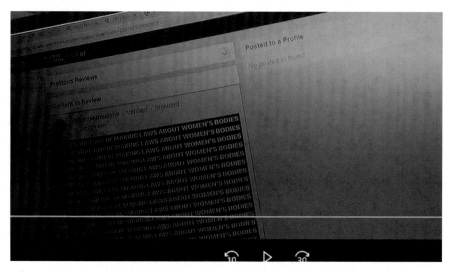

This is the phrase referenced in Picture 6 that Facebook gave an allowance for, despite its violating Tier 3 of the Hate Speech policy for exclusion. The phrase excludes all men from the political process (making laws).

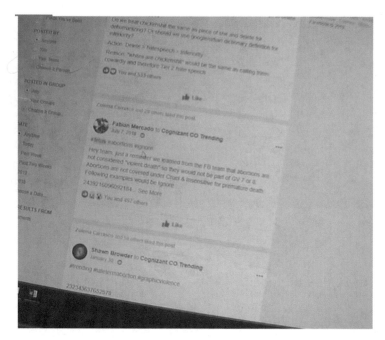

This is guidance given from the Facebook team telling us that abortions are not considered "violent death" under the Graphic Violence (GV) policy.

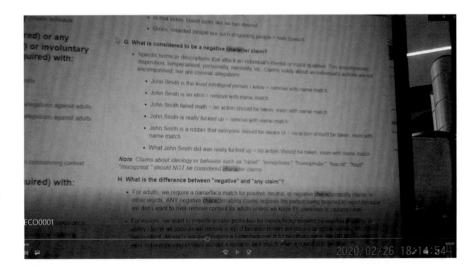

This screenshot shows under the Bullying policy in Known Question G the definition of a negative character claim, which is an attack on someone based on their character. However, calling someone racist, xenophobe, homophobe, fascist, or a Nazi is specifically excluded from this definition.

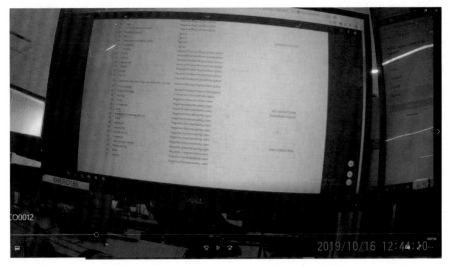

This shows the Bullying Slang List, a list created by Facebook of about a hundred common bullying terms we would come across often, and how to action them. The attacks "Trumphumper" or "Nazi" when reported by the user can never be deleted. However, the attack of calling someone a "snowflake" or "feminazi" was treated differently and would be taken down and deleted if reported by the user.

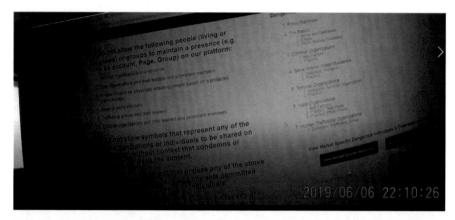

We see here the Dangerous Individuals and Organizations policy, which deals with terrorists, mass murderers, and hate groups. It is very strict as to what can be shared regarding these individuals or groups. In most cases, the only way to even mention anyone found in this policy is by condemning them.

Gavin McInnes, the cofounder of Vice Media and founder of the Proud Boys, is characterized as a hate figure in this list that includes individuals from the Nazi party, racist groups, and even Adolf Hitler. Since Gavin McInnes is on this list, any PSR (praise, support, or representing) of him is completely forbidden on social media. The only way to mention him on social media is in a condemning context.

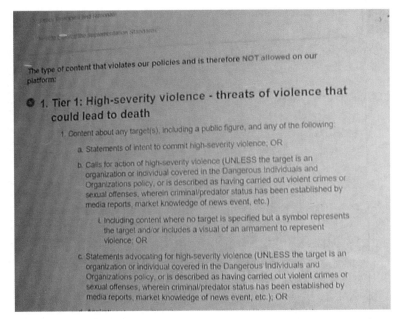

Under the Violence & Incitement policy (V & I), individuals from the Dangerous Individuals and Orgs policy are excluded from protection. So since Gavin McInnes is a hate figure on that list, I can say, "Let's kill Gavin McInnes," or any other variation of a threat of violence against him, except for a direct statement of intent to commit high-severity violence such as "I am going to kill Gavin McInnes." High-severity violence only covers stab, shoot, kill, etc. So I can still say, "I am going to beat up Gavin McInnes," and this would never be deleted off the Facebook/Instagram platforms.

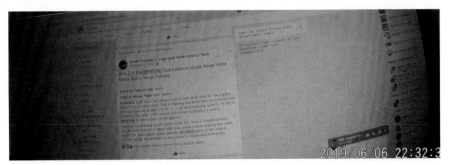

CNN host Don Lemon, while on national television, said that white men are the biggest terror threat in this country. Facebook acknowledges his statement violates their Hate Speech policy, yet they made a newsworthy exception to allow it.

Facebook's policy toward minor public figures would normally allow for an attack calling a famous minor "retarded." Facebook made a singular exception to their policy and scoured their own site for any instance of "gretarded" and had us delete them all.

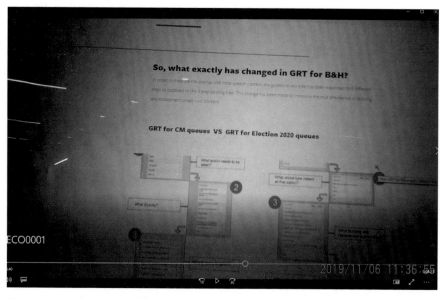

Facebook redesigned their entire content moderation queue in preparation for the 2020 election. The GRT is the guided review tree, and they wanted to be able to measure more accurately bullying and harassment within the context of the election.

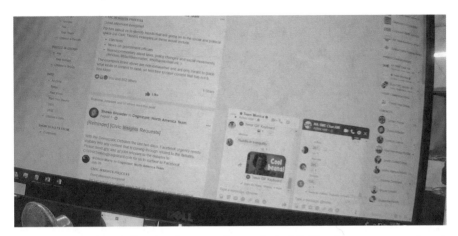

On August 1, 2019, Facebook urgently needed visibility into content related to the Democratic primary debates. We were told to document and communicate those examples to Facebook.

During the 2018 midterm elections, Facebook began a method of tracking voter fraud and other voter information. The action of "passing to SME" would send the content to Facebook for review. We would leave the note "VI" on any election-related content, which would allow Facebook to track those types of posts or comments throughout the platform.

Facebook gave us weekly recaps throughout the 2018 midterms, and here they counseled us on how to treat the phrase *monkeying things up*, which they told us to delete for hate speech, thus interpreting the statement in line with leftist media as a purposeful racist statement from gubernatorial candidate Ron DeSantis toward his competitor, Andrew Gillum.

Another weekly recap during the 2018 midterm elections was information about the Mueller Investigation. Facebook did a good job of making us aware of current political events to ensure we had context to correctly flag trends and delete violating content.

The initial guidance regarding Ciaramella on 11/6/2019 was to delete his name under the Privacy policy, with the implication he was undercover law enforcement and was being exposed. However, within hours, Facebook changed this guidance and deleted any mention of his name under another unrelated policy. They even state their clear purpose: "We are to protect him." One must ask, If a similar pro-Trump whistleblower were to go public, would they receive the same protections?

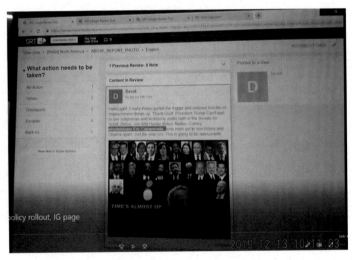

This is the type of post I was constantly deleting during November and December 2019. Any mention of Eric Ciaramella meant we instantly deleted the post, regardless of the context. Also in this image you can see the decision tree on the left, which allowed us to determine which action to take on the post/comment.

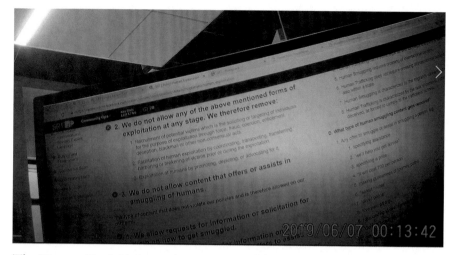

The Human Exploitation policy gave us guidance on human smuggling and human trafficking. Here we see that Facebook does NOT allow content that offers or assists in smuggling, but they DO allow requests for information or solicitation for help on HOW to get smuggled.

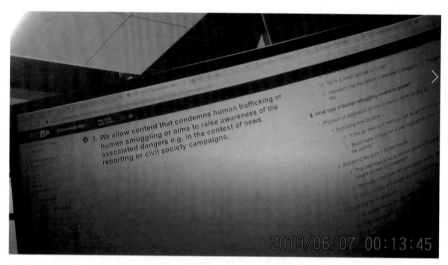

In Known Question E on the far right, you can see which type of human smuggling content is ignored. Discussing border patrol shift changes is allowed, as is explaining gaps in enforcement.

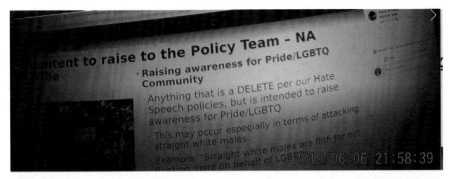

This is a screenshot of a training powerpoint in preparation for Pride Month, June 2018. We were told to raise examples of things that would normally be deleted, like attacks against straight white males, but that was given a second look because that type of hate speech raises awareness for Pride/LGBTQ.

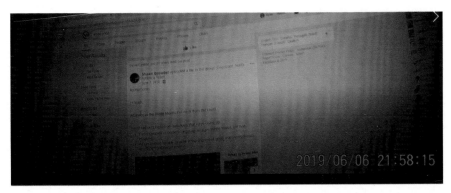

Facebook deliberated after we sent them examples of hate speech that were raising awareness for the LGBTQ community and decided that two examples of attacks against straight white males were non-violating.

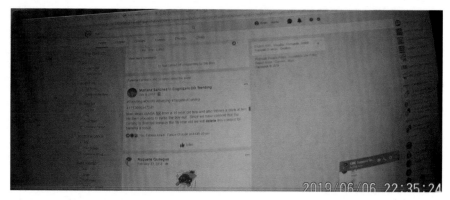

A viral video showed a Trump supporter being attacked, and Facebook decided to delete all instances of the video because of a minor infraction. In the video, the attacker curses at the minor. Many instances of the video had the cursewords bleeped out, but Facebook made no exceptions to allow any instances of this video, even those from news reports.

Hate Mail and Online Anonymity

Just recently a former VP of stats and info at ESPN named Ned Johnson[†] sent me a nasty private message on Twitter. His message said, "Delusional cultist. Please get help for your mental illness. Muted and blocked."

Technically this wouldn't violate Facebook's rules because I qualify as a public figure. If you do a Google[‡] online search of my name and there are at least five news articles about me in the last two years, then I'm a public figure, and most of the bullying policy doesn't apply. Calling me delusional is a negative character claim, but that only applies to private individuals. What I find most striking about Mr. Johnson's cyberbullying is that it's not anonymous. He uses his real name and title. You would think that he would have used a throwaway account to launch an attack on me. After he sent me this message, I proceeded to screenshot his private message and share it in a public tweet.

Generally, that is an assumption we make about cyberbullying—that it is more prevalent with anonymous identities. However, a 2016 journal article from *PLOS One* challenges the "popular assumption that online anonymity is one of the principal factors that promotes aggression."[5] This research shows that in the context of online firestorms, individuals are more likely to use their real names to make hateful or aggressive comments. Later in the journal article it talks about the benefits of being nonanonymous and making hateful or controversial comments:

> In contrast, the results support the idea that non-anonymous aggressive sanctions are more effective. Non-anonymity helps to gain recognition, increase one's persuasive power, and mobilizes followers. The result is also in line with public voices that observe an increasing social acceptance of non-anonymous digital hate speeches.[6]

I've seen this employed by social media influencers who quote someone who's attacking them and use that as a way to mobilize their followers. As if to say, "See, this person thinks I'm wrong, but I'm right," and this further solidifies the influencer's role as an "expert."

On February 10, 2021, US Senator Ted Cruz used a quote from Shakespeare's Act 5 of *Macbeth* to describe the impeachment hearing of former President Donald J. Trump[7], describing it as "full of sound and

---

† A pseudonym.
‡ We were told to specifically use a Google search to find news articles about public figures. No other search engine was recommended by Facebook.

fury, signifying nothing." In response, MSNBC's Andrea Mitchell tweeted "@SenTedCruz says #ImpeachmentTrial is like Shakespeare full of sound and fury signifying nothing. No, that's Faulkner."[8]

In this instance, Andrea Mitchell used her real account (nonanonymous) in an attempt to mobilize her followers and increase her persuasive power. This would not be possible with an anonymous account. This ploy, however, backfired because she was wrong. The quote indeed did come from Shakespeare, and she later apologized to Cruz.[9]

Female-Gendered Cursing

Another fascinating change made in 2019 was how Facebook treated female-gendered cursing against public figures. Prior to 2019, you could post comments directly under a famous person's page and call them a bitch, a cunt, etc. However, Facebook changed the rules and created a rule called purposeful exposure. So now if you tag Alyssa Milano or Melania Trump or post under their official social media accounts cursing at them with a female-gendered swear word, it gets deleted.

Mind you, this only applies to female-gendered cursing, not to male-gendered cursing. Somehow, for Facebook, calling someone a "bitch" is worse than calling them a "dick."

You'd think that for a company that claims they can't review every single comment, flagging millions of more posts a day (for female-gendered cursing) would be an issue. But apparently, it's not. The consequence of deleting more cursing is that human content moderators can see more trends related to these public figures and politics and share those trends with Facebook.

During an election cycle, Facebook can then look at these trends and attempt to shape the narrative or know which public figures are receiving more attacks. There are several other ways Facebook modified its bullying policy to advantage certain groups.

Claims of Ideology or Behavior

For example, in 2019, Facebook modified the bullying policy to allow attacks against ideologies or behaviors. The examples they gave here of allowed attacks are all words used to label and criticize right-wing or conservative individuals. This is not an exclusive list; however, it's striking that all of the examples they gave us to look out for are attacks against conservatives. These attacks are allowed and are conveniently marked as "claims about ideology or behavior." The bullying policy normally doesn't allow attacks against someone's temperament, disposition, mentality, or character. But attacks are allowed against conservative, Christian, right-wing Trump supporters.

> Note: claims about ideology or behavior such as racist, xenophobe, homo-phobe, fascist, nazi, misogynist should not be considered negative character claims.[10]

This means I can call you a "racist homophobe fascist" and the comment will never get deleted, even if you report it a million times. The net effect of this policy change allows more hate speech against right-leaning individuals.

Even left-leaning individuals can see the blatant injustice of this policy. It's like saying, "I'm going to let Person A punch Person B, but if Person B hits back, he's going to be arrested for fighting."

The left gets to call the right all sorts of names, but not so much the other way around.

Exaggeration of Speech

Facebook gave specific guidance to allow the phrase "Get Trump's dick out of your mouth." Normally, this would be a claim of sexual activity. For example, if I say, "John Smith had sex with Mary," this would get deleted no matter who reports it.

Facebook is saying that we shouldn't take this in its literal meaning, that we shouldn't interpret this as Trump literally having sex with a constituent and inserting his dick into their mouth. Now, this would apply to any politician. So technically I should be able to say, "Get Kamala Harris's dick out of your mouth." There is some overlap with the purposeful exposure policy here. However, if I posted under Kamala's official page and was advocating for her to engage in sexual activity with me, then that could be deleted.

Phrases like "suck my dick" are considered targeted cursing, so that would be allowed under Kamala's page.

But if I said "@KamalaHarris I want to have sex with you," that would be deleted.

What's telling is that they made these changes to allow more political discourse (if you consider putting a politician's dick in your mouth "political discourse") all while Trump was running the country. Why so much effort to modify the policy while Trump was up for reelection? We know there are many attacks against Trump supporters, and simultaneously Facebook was opening up the floodgates to allow more hate speech. This is no coincidence. We see the same modifications to the policy to help promote more attacks on police, which I explore in a later chapter.

Just like every other policy, it's obviously clear that Facebook's bullying policy has been weaponized and modified to conform to leftist ideology and to punish right-wing users of Facebook.

CHAPTER TEN

Facebook in Black and White

Do I really need to say I don't support any form of racism? I love meeting people from different cultures and think America is a great melting pot of cultures and that success is possible regardless of skin color, race, ethnicity, or culture. Call me a red, white, and blue patriot who tries my best to be color-blind.

As a content moderator, I often did see blatant racism that targeted people because of their pigmentation. I saw racism in every form, against whites, blacks, Asians, etc.

I also deleted slurs in Spanish that targeted nationalities. For example, calling a Peruvian a *"come palomas"* is something we would delete. At one point, we considered deleting *"Venacos"* as a reference to Venezuelans. But the policy team at Facebook decided against it. *"Venacos"* was still allowed as a way to refer to Venezuelans. We had a long list of slurs we could refer to that encompassed more than twenty countries in Central and South America.

When I transferred to the North America market in July 2019, I had to learn new slurs to delete. Obviously, the N word was deleted, along with "cracker" and a host of other words, some more obscure than others.

It was quite an education.

But if we're going to have a policy, shouldn't it apply to everybody equally?

* * *

What caught my attention was the double standard that applied to white vs. black. For example, I came across a post that was quoting Marcus Garvey. After researching his name, I discovered he was a black nationalist from the 1920s. This is a big part of our job, researching and knowing historical figures. I believe I asked for guidance on that job, and I also checked our Hate Figures List. He was not on the list, although there are many other nationalists on that list, many of them white nationalists, such as League of the South.

Marcus Garvey is now propped up as a hero for the Black Lives Matter movement. In fact, in a recent article they consider him a leader and someone to look up to: "Garvey and others assert that the only satisfactory and permanent solution to the problem of black-white relations is the separation of black people from the dominant white majority."[1] In many ways, this is similar to the belief of the Nation of Islam, which believes that communities need to develop and support their own and prohibit intermarriage or race mixing.[2]

Even in his time, Marcus Garvey was considered too extreme for black civil rights activists. He openly advocated for segregation and was considered a black nationalist.

Although his legacy as a leader and activist lives on, Garvey's separatist and Black Nationalist views were not embraced by many of his peers. In fact, W.E.B. Du Bois of the NAACP famously said, "Marcus Garvey is the most dangerous enemy of the Negro race in America and in the world."[3]

Similarly, another segregationist in Great Britain during the same time period was Oswald Mosley. Mosley is on our Hate Figure List. Later in Mosley's political career he did align with the fascist movement. But before leaving his position as an MP (Member of Parliament) with the Labour party, he was well regarded in the newspapers of the time during the 1920s. Mosley was staunchly against immigration from other countries and advocated for racial segregation, as well.

Both advocated for racial segregation, and their policies were racist in nature. Yet Facebook treats these individuals differently and prioritizes them differently. I realize that Mosley is more notorious and well known than Garvey. But in my opinion, this still shows the double standard with regard to advocates of Black Nationalism.

It seems that Facebook's Hate Figure List contains any and all White Nationalists, including contemporary leaders in Europe. However, Facebook doesn't seem to mind Black Nationalists doing the same thing—stoking the flames of resentment and pushing race as the denominating factor of one's nationality.

This seems to match the mainstream media's immediate response to shootings of black men and slow, lethargic coverage of white children being shot by blacks.[4] Color only matters if you're a black victim.

The 2016 Dallas Police Shooting

Within this same vein, Black Identity Extremism was one of the labels used to designate domestic terror threats as categorized by the FBI.[5] They since changed the name to "Racially or Ethnically Motivated Violent Extremism," to come across as more politically correct.

But it's clear this is also a threat, right along with certain White Nationalist groups like The Base. When five police officers were killed in Dallas on July 7, 2016, during a Black Lives Matter protest[6] (Price & Shadwick, 2016), this was a clear indicator that extremism is alive and well on both sides of the racial spectrum.

The 2016 Dallas shooter Micah Johnson told police that "he was upset about the recent police shootings" and "wanted to kill white people, especially white officers."[7] The New York Times characterized this mass shooting as "retaliatory violence" as payback for deaths of black men in police custody.[8]

Yet the media seems to ignore racially motivated killings when the attacker is black.

How Facebooks Treats Race

At Facebook, when I would search for the terms "White Pride" vs. "Black Pride," the results were bizarre. This was on our internal message board to give us guidance when actioning jobs and often directly referenced the policy or supplemented Facebook's policy.

There were many "call-outs" about racism when I searched "White pride." However, when I searched "Black pride," I found positive terms and topics involving the Black Lives Matter movement, which described "Black pride" as a positive thing. I acknowledge that the phrase "Black pride" is associated with the civil rights movement, and I respect that movement to fight against the systemic racism and segregation that existed in the middle of the twentieth century. I also acknowledge that although many steps have been taken to address racism, there are still many communities and individuals affected by bias and bigotry in our day and age.

However, I don't understand why "White pride" has to always be associated with a very small group of racists. Can't I take pride in the fact that our country is the only one in history that fought a war to end slavery? In our culture, we have been conditioned to believe that "White pride" refers to racist white people, whereas "Black pride" refers to people fighting for their civil rights. Linguistically there is no difference between these two phrases; however, the imbued meaning gives these two phrases opposite meanings. Can any race be proud of their race? Should all races be allowed to celebrate their culture?

Facebook's own Hate Speech policy allows for expressing pride in one's race as long as you don't put down another race. Unless you're white, then whatever you say is most likely a "hate slogan."

From Facebook policy:

Content supporting white pride is violating as it is a hate slogan, and has always been.

A violating example would be "I am white, I am proud, white pride!"[9]

Another coworker asked in the comment section why "White and Proud" is allowed, but "proudly white" is not allowed. That seems like a very nuanced distinction, and I don't know the answer, nor was I able to document the response that was given.

Here are some examples of what I found when I searched in SRT workplace* for "black power" vs. "white power":

Searched "black power" in SRT workplace
Results:
"Weekly Policy Clarification Recap
1. Is the slogan 'Black Power' considered a violation under hate speech? No, 'black power' is a slogan representing a movement in support of rights and political power for black people."[10]

Searched "white power" in SRT workplace
Results: first result is for white nationalism which is a a violating ideal.[11]

---

* SRT workplace is the internal company message board.

Facebook also has another list of "Hate Slogans" that contain obscure number combinations and phraseology that white supremacists use. Pepe the Frog is also considered racist with enough context, and while at Facebook I saw that the "okay" hand gesture symbol would also be considered violating with enough context.

Context is key while actioning jobs, and we would have to go through a very complex process to determine if something was violating.

White Trash

When I started working for Cognizant, the phrase "white trash" was initially deleted. Then it became acceptable in 2019. Facebook changed their policy, so you could now use the phrase "white trash" with no consequences. Now, I understand that "trailer trash" and "white trash" are somewhat synonymous.

But can you imagine a similar phrase of "Asian trash" or "black trash" or any other race being used and accepted? I even asked my supervisor Jason[†] about this change of policy on 12/11/2019:

White Appalachian Trailer Trash

Content on the screen that is being actioned upon:
"posted by Andrew: Sam Kim, lmfao, bro, chill with your victim reflex. I didn't say you were Japanese—that's just an example of you not being oppressed, you little prick. Nor did I say that any Asian can't understand poverty, only that sad, pathetic YOU, Sam, are clearly disinterested in poverty and matters of class—you probably think deep down (and this is partly the fault of our worthless curriculum) that **white Appalachian trailer trash** still has it better than poor, victimized Sam. It's not about being woke (if I ever use that word unironically about myself, please shoot me) it's about having a single little bitty ounce of compassion for your fellow human being that is not predicated on their physical/cultural similarity to you. If you had it, you wouldn't in the first place have taken to FMC to bitch about the vewwy vewwy mean MP staff, and you certainly wouldn't have dredged up this stupid thread again still somehow thinking you were in the right all along."

Ryan: Jason, in this job so, Sam's like hey people are bugging me because they think I'm Japanese and don't speak English or whatever. So he's saying Sam

. . . you probably think deep down that white Appalachian trailer trash still has it better than poor, victimized Sam . . .

Jason: White trash we don't do anything for anymore.

Ryan: So we ignore for that now?

Jason: There's only bullying there.

Ryan: Yeah I'm only seeing bullying. They're not really attacking him based on his PC [protected characteristic]. Cool. Thanks.[12]

It's important to remember that one of the main sources we used was Urban Dictionary. However, if I were to use Urban Dictionary to search for the term "white trash," I would find the following:

White Trash

A derrogatory [*sic*] term for very poor white people. A racist epithet, not because it specifies race, per se, but because it implies that unsophisticated, disadvantaged white people require special distinction as such to differentiate them from your average poor, huddled masses yearning to be free.

People described as "white trash" tend to be more overtly racist than most, because their very lives prove the fact that you don't have to be black or hispanic [*sic*] to be underpriveleged [*sic*]. They blame non-whites and reverse racism for their oppression . . . then vote for Bush.

There's trash . . . and then there's white trash. Some folks you ʿspect to be lazy good-fer-nuthin crooks. But white folks? It just ain't natural.[13]

Even according to the source we're supposed to use, the phrase "white trash" should violate Facebook's policy. It is a derogatory term for poor white people. It is attacking people based on their PC (protected characteristic), that of being white.

Tyler Edwards‡

---

‡ A pseudonym.

12 hrs[ago] 12/5/19 6:41pm

#Hatespeech #Trash #Filth

This post is to provide new and updated guidance for "trash" and "garbage" as well

as clarify slur + shit. This is an exception made as generally they are used as comparisons to Filth and not comparisons to inanimate objects. The call-out for inanimate objects only being deleted for designated attacks towards woman is still relevant this is only in conjunction with it to avoid any confusion.

- Men are trash → Delete > Comparison to Filth
- Women are garbage → Delete > Comparison to Filth
- women are pieces of trash → Delete > Designated Dehumanizing

Please use cultural context to establish if a term or phrase that uses trash/garbage but has a different cultural meaning

- You are white trash → No Action – The phrase "white trash" is used to identify someone as a person who is both financially poor and white. This only establishes a subset, not an attack.
- This is the most white trash thing I've ever seen → No Action

1. How would we treat slur + shit

A. In most instances, shit is being used as a synonym for stuff, so it would be a descriptor of the things the slur does, not a statement on the slur itself. When it is clear that slur + shit is being used as a replacement for slur + stuff action as slur

[emoji reactions by] 192[14]

At Facebook, they gave an exception to allow Don Lemon to say "white males are terror threats."[15] Can you imagine them allowing that same exception for any other race? Imagine Don Lemon saying "black males are terror threats" or "Hispanic males are terror threats." He would be immediately fired for such a thing. But somehow culturally it's okay to attack the white guy.§ As we saw in another chapter, Facebook also gave an exception to their policy to attack "straight white males as filth" for not supporting the LGBT movement.

White Lives Don't Matter

Facebook considered a White Lives Matter rally was racist, as shown in the following comment section from Facebook's internal message board, known as SRT Workplace.

> Here is the comment section about a White Lives Matter rally posted by Deeprek Hundai⁑ (no longer works there as of 2019):
>
> **Deeprek Hundai:** Hi all, I've gotten a question on if news videos (for example - CNN Coverage of the event) would count as a livestream representation. Since the intent of news streams is to share information with the public, I'm going to give interim guidance right now to NOT take down.
>
> Mainstream news reports. News streams (for example CNN/Buzzfeed coverage) would NOT count as PSR for the hate event, please do not delete mainstream news coverage. If you have any edge cases please escalate them within this threat. Thanks![16]

White Guilt

Every single exception I saw regarding race was to allow more attacks against white people. This is reminiscent of popular culture, as we saw the popular Netflix show *Dear White People*, which was a lecture to white people and their inherent problems. We see this concept being taught in universities across the nation, the idea that whites should recognize their whiteness and acknowledge their white privilege and white guilt.[17]

From this same article by Monnica T. Williams, PhD, she defines black pride as "pride in the accomplishments and resilience of a racialized group in the face of continual oppression." Whereas white pride is defined as "continuing to accept unjust benefits that ultimately come at the expense of others."[18]

_____

§ See Chapter 13 for the full quotation of the guidance regarding Don Lemon's mention of white males as terror threats.
⁑ A pseudonym.

My biggest issue is with the concept of white guilt. This is encouraging people to feel bad because of their race, or about what their race has done in the past. I believe people should be responsible for their actions and not the actions of others. There is a Bible verse that talks about the sins of our ancestors, which I don't necessarily agree with.

Deuteronomy 5:9 says, "Thou shalt now bow down thyself unto them, nor serve them: for I the LORD thy God am a jealous God, visiting the iniquity of the fathers upon the children unto the third and fourth generation of them that hate me."[19] From this passage, it seems as if God were in favor of collective and generational guilt.

And yet there is a conflicting verse later in the same book of Deuteronomy. Deuteronomy 24:16 says, "The fathers shall not be put to death for the children, neither shall the children be put to death for the fathers: every man shall be put to death for his own sin."[20]

In conclusion, I don't think I should be held morally responsible for something my grandparents did, or if not in my family, then by people who share my skin color at a different period of time. The great thing about America is that we celebrate individual contributions. I do think we should be responsible for our own actions.

However, I do wish the "woke" social justice warriors would clarify for us just how long we should be punished for our ancestors' actions. God stopped at four generations. It seems the social justice warriors want to beat that record.

CHAPTER ELEVEN

Facebook's Treatment of Antifa, Police Officers, and Drug Cartels

Facebook's community standards are designed to prevent real-world violence. There are multiple sections of the policy that deal with violent acts. Violence & Incitement (V & I) deals with direct threats to harm individuals and specific buildings, property, etc. "Coordinating Harm" deals with more general statements of committing a crime or causing property damage, even includes if you're speaking positively of violence or property damage, theft, etc. It also includes harm against property and people, and voter interference. Here is most of the Coordinating Harm policy, verbatim:

2. Harm against property

   1. Coordinating (statements of intent, calls to action, representing, supporting or advocacy) OR depicting, admitting to, or promoting the following acts committed by you or your associates.
      1. Vandalism
      2. Hacking when the intent is to hijack a domain, corrupt or disrupt cyber systems, seek ransoms, or gain unauthorized access to data systems.
   2. Coordinating (statements of intent, calls to action, representing, supporting or advocacy) OR depicting, admitting to theft when committed by you or your associates, as well as speaking positively about theft when committed by a stranger.

3. Harm against people

    1. Depicting, admitting to or promoting the following when committed by you or your associates.

        1. Acts of physical harm against humans, including acts of domestic violence EXCEPT when shared in a context of redemption or defense of self or another human.

        2. Coordinating (statements of intent, calls to action, representing, supporting or advocacy) OR depicting, admitting to, or speaking positively about swatting when committed by you or your associates.

        3. Content that depicts, promotes, advocates for or encourages participation in a high risk viral challenge (NOTE: Escalate to Content Policy all viral challenges which have not been previously designated).

4. Voter Interference

    1. Voting interference through physical/verbal acts and misrepresentation of operational/procedural details include:

        1. Offers to buy or sell votes with cash or gifts, and methods for voting or voter registration.

        2. Statements that advocate, provide instructions, or show explicit intent to illegally participate in a voting process.

        3. Misrepresentation of the dates, locations, and times, and methods for voting or voter registration.

        4. Misrepresentation of who can vote, qualifications for voting, whether a vote will be counted, and what information and/or materials must be provided in order to vote.

    2. Census interference through physical/verbal acts and misrepresentation of operational/procedural details include:

        1. Statements that advocate, provide instructions, or show explicit intent to illegally participate in a census processes, methods for voting or voter registration.[1]

Under the Violence and Incitement policy, there's also a section that says that it's against Facebook rules to advocate for violence due to the outcome of an election. With all of these rules against advocating for violence and property damage, however, we still see massive organizing for riots and protests. Where is Facebook in all of this?

Why is Facebook not shutting down more pages from Antifa and other groups who are planning mass riots? Why do they not even address the question? And why does the mainstream media let them get away with not reporting on it?

I think Facebook is turning a blind eye by defining these riots as protests instead of what they are—opportunities to destroy cities and businesses.

Just from a simple search on my Facebook for Antifa, I found two violating posts. One is advocating giving Trump to Iran, who placed a bounty on him in January. The other is calling for the overthrow of the government.

Facebook's policies are not being applied to the thousands of protests that have happened since George Floyd's death in May 2020. For example, if I post a video of a rioter stealing something from a store and I say, "That's awesome," that would be deleted. However, with regard to vandalism, I have to be personally involved or it has to be my associates. But if I show a video of vandalism with the caption "My friends and I destroyed this car," only then would it be deleted. Facebook allows me to post videos or photos of vandalism and speak positively about it, as long as I don't say that it was committed by me or my associates.

The third section of the policy that Facebook could use is Dangerous Individuals and Organizations, which we commonly referred to as Dangerous Orgs. As we saw in Chapter 8, they include criminal organizations within this policy. In this post from 2017, you can see that Facebook doesn't consider Antifa as a hate organization:

Cynthia Tomlinson*
November 3, 2017

#HighPri #Nov4Protest

On Saturday Nov. 4th, there will be a planned nationwide protest titled "The Trump/Pence Regime Must GO", against President Trump and Vice President Pence. This protest is occurring in at least 9 cities. This protest is expected to span a couple of days (similar to the Occupy Wall Street protest). There have been unconfirmed links between this protest and Antifa. There will potentially be counter protesters and it might become violent. The protest

---

* A pseudonym.

will take place in the following cities: Atlanta, Austin, Boston, Chicago, Los
Angeles, New York, Omaha, San Francisco and Seattle.

Important Call-Outs:

Antifa is not considered a hate org.

Look out for the following violations: coordinating harm and hate speech.

Share trends and pose questions in the comments below.

Read More: https://www.newsweek.com/antifa-rallies-november-4-promise-
remove-trump-white-house-700406

[liked] You and 494 others   Seen by 964[2]

A comment below this post said the following:

[comment by Jack Cornelius[†]]
Jack Cornelius: Also remember to check counter protesting groups against the
designated hate organizations list. Given the recent Unite the Right demon-
strations in Charlottesville and elsewhere, it's highly possibly that there may
be white-supremacy groups doing counter-protesting. Always be on your feet,
you never know what may happen. [liked by 15][3]

Based on new undercover journalism videos from Project Veritas,[4] we know
that Antifa trains to commit violence. And based on video evidence at
numerous protests, we know that they engage in widespread looting and
destruction of property.

Facebook is quick to add groups like Proud Boys to their hate list, and
individuals like Kyle Rittenhouse are rapidly (within days) labeled as mass
murderers under this same policy.[‡] The mass murder in New Zealand in
Christchurch was a tragedy, and out of that arose a brand-new section of
the policy for Facebook.

But when massive organized riots occur in dozens of cities, Facebook
is quick to cover for Antifa and say, "Nothing to see here." If Facebook was
genuinely concerned about violence, they would have their AI search for

† A pseudonym.
‡ See Chapter 15 for a more thorough analysis of Kyle Rittenhouse.

any support of vandalism in US cities and delete it. If they can delete any instance of attacks on Greta Thunberg (see Chapter 5), they can find posts supporting vandalism. Any excuse that Facebook doesn't have the man-power or resources is simply a lie.

Facebook loved it when I brought up trends about "Boogaloo" and "Civil War"[§] and were very aware of a peaceful gun rally in Virginia. But when simultaneous protests with looting and rioting occur, they support BLM and Antifa. Facebook itself should be considered a terrorist organiza-tion, for allowing groups like Antifa to exist within the platform. This tran-script from guidance given to content moderators is very telling. In it, both Antifa and Proud Boys are mentioned, but only one of them is considered a dangerous organization:

Video 7

Hank Johnson[¶]
August 13th 2019 at 3:29pm

[Heads up][Proud Boys/Antifa Rally]

Event Description: Proud Boys and Antifa to hold a rally against each other in Portland this Saturday.

Policy Guidance: Proud Boys are internally classified as a hate organization. All content

praising, supporting and representing this organization should be removed from our platform.

If you have any questions, please feel free to reach out.

[liked] You and 494 others   Seen by 964[5]

---

§ See Chapter 15 for more discussion regarding "Boogaloo" and Facebook's interest in those types of trends, along with information about the peaceful gun rally in Virginia.
¶ A pseudonym.

The War against Police

I never would have believed we'd come to the day in American history when cities would voluntarily destroy their police force. Yet it's happening in many cities across the country, including Minneapolis, Minnesota. The city council chose to defund the police force, with devastating consequences for the residents.[6]

I believe this is all part of a strategy that's been in play for at least ten years to weaken our nation. Local police are the one thing standing in the way of anarchy and totalitarianism. We saw during the Obama administration an uptick in the vilification of police and their use of force. We saw the rise in popularity of Michael Brown, who in 2014 was shot by police after robbing a convenience store. The misleading slogan "hands up, don't shoot" originated from that incident, where Michael Brown charged toward an officer and ended up getting shot. The Obama administration leaped to the conclusion that the officer was guilty before the blood had even dried, even though a later investigation by the Obama Justice Department cleared the officer of any wrongdoing.[7]

Attitudes toward law enforcement have never been worse in our country. Officers now regularly fear being lynched by the mob of popular opinion, and of the prosecutor siding with the mob. Recently, a grand jury decided on the case of Breonna Taylor, who in March 2020 was killed while a search warrant was served for her at her residence. I tried to have a civil discourse on Twitter (I know that's not very likely), and this ended up happening:

> On Twitter I sent out "Okay, here's the police warrant, the police report (which is pretty much blank), and the lawsuit from her family. There was a warrant and Breonna Taylor is listed. I'm just gonna leave it at that and let you all research it yourselves." In response, @TJMair responded by saying "Speaking of actual extremists, after you followed me (which is odd), I checked your profile, @realryanhartwig. You're proudly part of the far-right group, Project Veritas. There's a good chance that you're a Russian bot or a sock puppet of James O'Keefe's!"[8]

Earlier I asked this individual if there was a warrant for Breonna Taylor, and he said that there wasn't. I tweeted out the link to the search warrant, and then he replied by attacking me based on my affiliation with Project Veritas and calling me a puppet of James O' Keefe. We can see that there is a huge misinformation campaign regarding some of these recent shootings: George Floyd, Breonna Taylor, Michael Brown, the list goes on and on.

Yes, these are tragedies, but by and large these individuals were not complying with verbal orders from police and/or were under the influence of drugs or alcohol.

I've worked as a security guard off and on for the last six years. I've worked at the Super Bowl, grocery stores, bars, and golf tournaments. I've had the chance to interact with cops on many occasions.

I remember one incident in Avondale, Arizona, where I dealt with an individual who was clearly on some kind of drug. This example illustrates the unpredictability that cops deal with on a daily basis. I wrote about this in my book *30 Jobs in 30 Years: A Millennial's Guide to the 21st Century*, on page 123:

> My second favorite experience also occurred with the Avondale PD. The cashier at the deli, whose confidence I had won over with my fluent Spanish, told me that a man had stolen a sandwich. I approached him at the door. He convinced me for the moment, but I was still suspicious. He seemed to be exhibiting strange behavior, and so I followed him at a distance as he mozied around the sidewalk outside the front of the store.
>
> He eventually ended up behind the Fry's store by the wooden pallets. After observing him, it seemed very obvious to me that he was on some hard drugs. I called PD, hoping he would stay in the area until they arrived. PD arrived and had him spread his legs and cuffed him. As the officer was attempting to put him in the back seat, he struggled and kicked. The officer wanted to make sure we had video documentation that he indeed had stolen. So I went upstairs, had the manager look at the DVR with me, and recorded the recording with my cell phone. Sure enough, he had grabbed a sandwich at the deli, told them he'd pay for it up front, then swiped a soda before bypassing the point of sale.
>
> I can only imagine the harm this individual could have done if he had possessed any kind of a weapon. It's situations like this that really put things into perspective. The majority of U.S. citizens don't interact with the police very often at all. They don't know what cops go through on a daily basis.[9]

How Is Facebook Involved in the War against Police?

In Fall 2019, Facebook allowed more attacks against police. Facebook's bullying policy protects private individuals. The only people not protected as private individuals are those considered public figures. You can qualify as a public figure one of three ways: hold an elected position, have more than a hundred thousand followers on any type of social media, or, alternatively, if your name is searched on Google and your name comes up five times in the last two years, then you are considered a public figure.

Cops don't fit any of the criteria for being a public figure, unless you're someone famous such as Sheriff Joe Arpaio from Arizona or Sheriff David Clarke from Wisconsin. For example, if I took a photo of a cop while he's on duty, by Facebook's definition that person qualifies as a private individual and would receive all the same bullying protections as anyone else.

For example, if I have a photo of a cop and call that cop a "cunt," that would be deleted for female-gendered cursing, regardless of who reports it. If I call for that cop's death, that would be deleted. If I threaten to release the cop's private address or phone number, that would be deleted under Section 1.7 of the Bullying policy.

However, under Sections 5 & 7 of the Bullying policy, some attacks on private individuals require that the person being attacked report the comment directly in order for it to be removed. This is what we call a name/face match.

For example, if I call Jack Jones a "good-for-nothing loser," this is considered a negative character claim. Under Section 5.4, this requires that Jack report the comment for it to be deleted. Otherwise the comment stays up.

Attacks that require a name/face match are considered less severe attacks. But if I call anyone a "whore," its gets taken down regardless of who reports it. But if I say "I despise Jack Jones," Jack would have to report it for the comment to get deleted.

If I were to call "Jack Jones," who is a private individual, an animal that is considered culturally inferior, that would always be deleted regardless of who reports it. "Jack Jones, you're a swine" would get deleted even if Jack didn't report it. "Jack Jones, you're a roach" gets deleted even if somebody else reported it.

If it's not an animal that's considered culturally inferior, such as a tiger or lion, that would require a name/face match.

This is shown in the Bullying policy under Section 7.5. On September 27, 2019, there was a post by Thad Franklin,** who posted guidance from Facebook about how we treat the attack "cops are pigs":

> "Thad Franklin to Cognizant: North America Team
> September 27 (9/27/2019 10:02pm)
> Parent Object: Police officer in uniform

_____
** A pseudonym.

Reported comment: all cops are pigs

Would we deleted to err on the side of the private individual depicted in the parent object and delete?

Since the update to KQ Y, the process/policy team at FB has given guidance that we should consider "pig" when targeting police as targeted cursing, which would require a NFM in order to delete. That is because of how the term is used in the NA [North American] market.[10]

To summarize the policy about cops, here is my own analysis:

- The bullying policy disallows comparing any private individual to an animal considered culturally inferior. This does not require the person in the image to report it. Any person identified by name or photo that is compared to an animal is a violation of policy.
- An exception to this policy was given on 9/27/2019, thus calling cops pigs would be an ignore, unless the cop in the image reports the post. The only explanation given by Facebook was "That is because of how the term is used in the NA market."
- Calling someone a pig isn't targeted cursing, it's calling them an animal. Facebook changed the definition of an attack in order to fit their own bias.

This post in SRT Workplace states that calling cops who are depicted in a picture "pigs" is no longer a delete. Previously, we would delete these types of jobs, because the police officers depicted are private individuals and they are comparing them to animals, which doesn't require a face match (NFM=no face match). Changing "pig" to targeted cursing now requires a name/face match, and it's very unlikely for the cop to find and report a comment calling him a pig. The rationale for this change doesn't make sense. The only reasoning they give is "because of how the term is used in the NA market."

This policy would allow me to take a picture of a cop on patrol and post it to Facebook with the caption "This cop is a pig," and it would stay up forever. It could get shared thousands of times and the ONLY way it would get deleted is if that same cop reported the thousands of instances of that image and caption.

Facebook's rationale for this decision is LITERALLY "because of how the terms is used in the NA [North American] market." That is the most insane response ever.

They literally made an exception to their policy to allow cops to be called pigs. It is despicable and is a slap in the face to every officer in the United States who risks his/her life on a daily basis to protect the citizenry.

The DOJ should investigate Facebook for instigating hate against police officers.

CHAPTER TWELVE

Kaitlin Bennett, the AR-15 Girl

Kaitlin Bennett, also known as the AR-15 girl, is a conservative gun rights activist who is often seen attending protests carrying an AR-15. She is very provocative, and every time she shows up at a college campus, the leftist mob is guaranteed to respond in force. Kaitlin describes herself as "the most hated conservative woman on the Internet."[1] She became famous in 2018 when she graduated from Kent State University and on graduation day carried an AR-15 rifle slung over her shoulder.

During my time at Facebook, we often had to research individuals to see if they are a public figure. If I didn't recognize Kaitlin, I would first have to find her name. While at work I would use Google to search "AR-15 girl." (They specifically instructed us to use Google.) The first search result was an article from patch.com describing Kaitlin Bennett with the headline "Come and Take It." To determine if she is a public figure, I would then search the name "Kaitlin Bennett" in the Google news tab.

As you'll see later, the issue of whether she is a public figure or private individual is extremely important. After doing a news search for her name, I easily saw she had five or more articles in the last two years. In fact, she had five news articles just from the previous week.

Also, I can search using Social Blade (https://socialblade.com/), which many content moderators used. We were not allowed to directly visit Instagram or Facebook while at work to check how many followers an individual has. Remember the rule that they are a public figure if they have a hundred thousand followers on any social media platform OR if they show up in a Google news search at least five times in the last two years.

Also, other people who are considered public figures are journalists, heads of state, or have some form of notoriety.

In the process of researching this book, I did a Google news search on October 20, 2020, for Kaitlin Bennett, and the fourth article down was titled "Kaitlin Bennett bombarded with 'Sh**t your pants' chant on college campus."[2] This is similar to the type of research I would have done while at Cognizant/Facebook.

I came across this meme very frequently between July 2019 and February 2020 in the North America content queue. The meme shows a photo of a real person face-down, passed out with the girl's dress pulled up about half-way, partially covering the buttocks.

There was a large amount of feces situated between the female's legs. I know, gross.

It seems that the female is passed out in some type of bathroom or dorm room. As content moderators, we had to analyze every aspect of this photo.

Was this really Kaitlin Bennett? This is key because if it was NOT Kaitlin Bennett, then we could easily delete the image for being a creep shot under the SEA (Sexual Exploitation of Adults) policy.

> SEA 2.3.1 "Secretly taken non-commercial imagery mocking, sexualizing or exposing the PDITI, aka 'creep shots' of real adults with: 1. Depiction of 'upskirts' OR 2. Depiction of sexual activity OR 3. Depiction of a commonly sexualized body part (breasts, groin, butt, and/or upper leg/thigh) OR 4. Any of the above with captions OR overlay text OR digital alteration OR audio which mocks, sexualizes or exposes the PDITI"

> Note: PDITI means Person Depicting In The Image.

> The Operational Guidelines section of the Sexual Exploitation of Adults policy describes Revenge Porn and Creep Shots.

> "H. What is considered a 'private setting' within our NCII/Revenge Porn policy?

> 1. While a public setting would be a space that is open and accessible to the public (e.g., roads, parks, beaches, town/city squares, etc.), we would consider a private setting to be somewhere where you would not expect many members of the public to be at one time. E.g,

1. a setting where it is not possible determine if the imagery was taken in a public setting (e.g., close-up of genitalia).
2. a setting that is enclosed and includes one or few people (e.g., anywhere in a residential building, apartment, car, hotel room, bathrooms, office, bedroom).
3. a setting that may be outdoors but is 'private' in nature due to the limited number of people in the imagery or due to it's seclusion (e.g., people engaged in a sexual act within a car parked on a public street).

Note: Please disregard photos that have been produced for media or pornography (e.g., for Playboy magazine, etc.)."

"J. What is the scope of 'vengeful context' for content to be considered revenge porn/NCII?

1. When the intent to share sexually explicit material is to humiliate/intimidate the PDITI or to cause them distress or embarrassment. Intent can be 1) explicit (accompanying vengeful text or captions or inherently vengeful words or phrases) or 2) implicit (posting the material in a dedicated space created for the purpose of Revenge Porn/NCII such as a group or page with a vengeful title)."[3]

To summarize this, Facebook policy doesn't allow you take a photo of someone without their permission and post the photo if it shows someone in a sexual situation (e.g., skirt pulled up). And Facebook doesn't allow revenge porn, but there has to be enough context to show that it's revenge porn.

In addition to the SEA/Revenge Porn policy, Facebook also knew that college coeds would take embarrassing photos of one another. A completely separate policy known as the Bullying policy also covers these types of photos. Bullying policy Section 4.6 states what is not allowed: "Target private individuals or involuntary minor public figures with: depiction of others in the process of, or right after menstruating, urinating, vomiting, or defecating where context further degrades the individual or contains an expression of disgust." If you hover over the phrase "further degrades," it defines "further degrades" as:

- Laughter ("hahaha")
- Negative character or ability claims
- Positive reinforcement ("he deserved that")

- Encouraging the broadening of distribution/audience (e.g., Like, Share, Tag, etc.)
- Mocking[4]

Therefore, if you draw a clown face on your college buddy with a Sharpie, and share it on Facebook with the caption "what a dork," that would get deleted by content moderators. But if your college buddy's name is Ben Shapiro or Alyssa Milano, then we would have to leave it up, since they are public figures and that section of the Bullying policy does not apply to public figures.

It's important to know that for the Sexual Exploitation of Adults policy, it doesn't matter if you're a public figure or a private individual. The policy applies to everyone equally regardless of status or popularity. And the SEA policy is higher in the hierarchy than Bullying, so if we find it meets the criteria for SEA, we wouldn't even have to worry about a Bullying violation.

However, the Bullying policy does distinguish between public figures and private individuals.

We already illustrated that Kaitlin Bennett is a public figure, BUT we don't yet know if the girl in the now-famous meme is actually Kaitlin Bennett. Here is what we discussed.

Chris Thomas* to Cognizant CO Trending
November 2, 2019
#trending #doitforstate #kaitlinbennett #defecation

SRT ID: 746334342438681

The photo above is being circulated in different queues, a resurface of the rumor passed around earlier this year that it is a photo of Kaitlin Bennett defecating on herself at a party. There is nothing out there debunking the statement currently and many jobs contain the same context, that it is her on the ground. As she is considered a public figure, she would not be covered under that portion of bullying and no action should be taken on this content.

No action > benign

[reaction] You and 426 others[5]

---

* A pseudonym.

Saying "there is nothing debunking the statement" means there is ambiguity as to whether it's Kaitlin Bennett. However, Facebook's own policy in Known Questions tells us that if we're unsure whether the person is a private individual or a public figure, we should default to treating them as a private individual.

Facebook policy

"Implementation Standards > Bullying > Operational Guidelines > Identifying a Target

What if I can't fully see the face of a person to determine if they're a public figure?

You should default to treating the person as a private individual. You should only treat people as a public figure if you can establish that they are a public figure from the content.

How do I treat an attack towards a depiction of a public figure AND a private individual when I am unsure who is being targeted?

If you cannot establish who the target is from context, tend towards protecting the private individual.[6]

Although the female in the image has blonde hair and her hair style is similar to that of Kaitlin Bennett's, she could be one of millions of women with a similar appearance. And although we don't know that person's name, their feces-covered photo is being used to ridicule Kaitlin Bennett.

One comment that I reviewed while at Facebook was from Lisa,[†] who said, "Go shit your pants again, Kait." Jessica[‡] replied to Lisa and said, "Lisa, I've seen the picture and I honestly didn't know a human could produce that much shit (laughing emoji). makes sense tho bc she is full of shit. I'm waiting for the day she gets her shit rocked. Who the fuck goes to a place where ELEMENTARY SCHOOLERS DIED in a school shooting to ADVOCATE FOR LESS GUN LAWS. That's a special type of fucked up."[7]

This comment brings into question yet another section of the policy, technically known as the IS (Implementation Standards). Facebook loves their acronyms.

The Violence and Incitement Policy (V & I) prohibits comments and posts that could lead to real-world violence. There are three tiers. Tier 1 deals

[†] A pseudonym.
[‡] A pseudonym.

with high-severity violence (e.g., "I will cut you"), Tier 2 deals with mid-severity violence (e.g., "I will beat you up"), and Tier 3 deals with anything lower than a punch in severity (e.g., slap, shove, drag, spit).

Tier 1: High Severity violence—threats of violence that could lead to death

1. Content about any target(s), including a public figure, and any of the following: . . .[8]

Tier 2: Midseverity violence—threats of violence that lead to serious injury

1. Content about private individuals, minor public figures, high-risk persons, or high-risk groups AND any of the following:

   a. Statements of intent to commit violence; OR
   b. Statements advocating violence (UNLESS the target is an organization or individual covered in the Dangerous Individuals and Organizations policy, or is described as having carried out violent crimes or sexual offenses, wherein criminal/predator status has been established by media reports, market knowledge of news events, etc.); OR
   c. Calls for actions of violence (UNLESS the target is an organization or individual covered in the Dangerous Individuals and Organizations policy, or is described as having carried out violent crimes or sexual offenses, wherein criminal/predator status has been established by media reports, market knowledge of news events, etc.).
      i. including content where no target is specified but a symbol represents the target OR
   d. Aspirational or conditional statements to commit violence (UNLESS the target is an organization or individual covered in the Dangerous Individuals and Organizations policy, or is described as having carried out violent crimes or sexual offenses, wherein criminal/predator status has been established by media reports, market knowledge of news events, etc.).

Notice that Tier 1 is more expansive and includes content about "any target," whereas Tier 2 only applies to "private individuals, minor public figures, high-risk persons, or high-risk groups." Tier 2 does not give protections to "public figures." It does, however, protect high-risk people and groups.

As a quick aside, you also see this policy does NOT protect people covered in the Dangerous Individuals and Organizations policy, including people like Tommy Robinson, who protested Muslim rape gangs in England, or Gavin McInnes, a comedian and founder of the Proud Boys.

"High-risk persons are defined by Facebook under Known Question (KQ) Section I.

I. Who is considered a high-risk person?

1. Heads of state
2. Former head [sic] of state
3. Next in line for heads of state
4. Candidates for head of state
5. Candidates running for nationally elected positions (for up to 30 days after election if candidate is not elected to position)
6. Nationally elected officials
7. Former candidates for head of state
8. Specific law enforcement officers
9. Witnesses and informants
10. People with a history of assassination attempts
11. People listed as targets on Hit Lists created by Banned Dangerous Orgs
12. Activists and Journalists[9]

As you can see, activists and journalists are high-risk people; therefore, they receive more protection under the Violence & Incitement policy. Per Facebook policy, I cannot say, "I hope Katie Couric gets punched," because she is a high-risk person (journalist). But I CAN say, "I hope Justin Bieber gets punched," because he is NOT a high-risk person, and he is neither a nationally elected official nor an activist.

However, Facebook's definition of "activist" is vague. They don't have a list of "activists," and that's why I emailed the Cognizant trending team (COPhxtrending@cognizant.com) and asked them if Kaitlin Bennett is considered an activist.

I emailed this question on February 18, 2020:

866924703780949

Delete V & I –mid severity
Is Kaitlin Bennett considered a high-risk person?
KQ I. high-risk person
12. Activists and Journalists

Best,
Ryan Hartwig
Process Executive-North America
Technical Ops- Digital Operations
"There are far, far better things ahead than any we leave behind" -C.S. Lewis
Cognizant
[address redacted]
6th Floor
Phoenix, AZ 85021
ryan.hartwig@cognizant.com[10]

The job I was referencing said, "I'm waiting for the day she [Kaitlin Bennett] gets her shit rocked. Who the fuck goes to a place where ELEMENTARY SCHOOLERS DIED in a school shooting to ADVOCATE FOR LESS GUN LAWS. That's a special type of fucked up."

Does the aspirational statement of hoping that someone gets their shit rocked qualify as midseverity violence?

Known Question (KQ) F. states

What is considered midseverity violence?
The equivalent of a punch or higher

- Including but not limited to
  1. Punch
  2. Kick
  3. Gag
  4. Hit
  5. Beat

- Does not include anything like [these are likely low-severity violence]
  1. Pinch
  2. Push
  3. Shove

4. Drag
5. Slap
6. Spit[11]

Next, I searched in Urban Dictionary (which Facebook told us to use) and found the term "shitrocked" defined as "a shortening of the phrase He/She got His/Her shit rocked. Meaning to get hit amazingly hard, or to be beaten bad in a sport or fight. Did you see that kid get hit? He got shitrocked! #rocked #shit #shitrocked #owned #pwned."[12]

Getting "beaten bad" and getting in a fight is very similar to getting "beat," so the phrase "I'm waiting for the day she [Kaitlin Bennett] gets her shit rocked" qualifies as midseverity violence. The only remaining question is whether Kaitlin Bennett is considered an activist. If she IS an activist, I delete the post. If she is NOT an activist, the post stays up.

This is a great example of how a simple meme of a passed-out college student becomes a puzzle involving legalese, research, and complex word analysis.

I don't believe I ever got a response to my specific email asking if she was an activist or not. And with regard to the feces meme, the guidance to allow it was never revoked, at least while I worked there. So young women can be "shit-rocked" on Facebook if they're conservative, and they won't be protected. Do you think that's encouraging or discouraging young women to become or from becoming conservative activists?

Chris Thomas[§] to Cognizant CO Trending
November 2, 2019
#trending #doitforstate #kaitlinbennett #defecation

SRT ID: 746334342438681

The photo above is being circulated in different queues, a resurface of the rumor passed around earlier this year that it is a photo of Kaitlin Bennett defecating on herself at a party. There is nothing out there debunking the statement currently and many jobs contain the same context, that it is her on the ground. As she is considered a public figure, she would not be covered under that portion of bullying and no action should be taken on this content.

---

§ A pseudonym.

No action > benign

[reaction] You and 426 others[13]

Between the feces meme and the "shit rocked" attack against Kaitlin Bennett, we had to research three separate policies, the Bullying and Harassment policy (B & H), the Sexual Exploitation of Adults policy (SEA), and the Violence and Incitement policy (V & I).

I'm sure you'll never look at a meme the same way, and I'm sure Kaitlin Bennett would be curious to know why Facebook allowed these attacks against her. This also illustrates how Facebook let hateful content thrive when the target was a conservative activist.

CHAPTER THIRTEEN

Newsworthy Exceptions: How Facebook Gives Exceptions to Its Policies, Including Hate Speech, When It Suits Them

On Wednesday October 28, 2020, Mark Zuckerberg and other tech CEOs including Jack Dorsey and Sundar Pichai were questioned during a Senate Hearing regarding their role in influencing the election.

At one point, Senator Markey from Massachusetts stated that anticonservative bias is a myth, and Zuckerberg told Senator Markey that Facebook doesn't make exceptions.[1]

During this hearing, Jack Dorsey, the CEO of Twitter, was asked whether or not Twitter has the ability to influence the election. He said no.[2] This is a laughable response, because it is glaringly obvious the immense power tech companies wield. Even if they didn't abuse their power as they're doing, they clearly have the potential to control the flow of information on a grand scale.

* * *

One of the easiest ways for Facebook to influence content and elections is by making "exceptions" to their own policy. They can do this by explicitly labeling certain jobs as "newsworthy exceptions" or by simply giving guidance on specific jobs that flies in the face of the wording of the policy.

We see this with the Kaitlin Bennett example, where the guidance clearly didn't align with the wording of the policy and we left up a photo presumably of a private individual passed out with her skirt pulled up, and feces coming out.

Another example was when more attacks against cops were allowed by permitting the phrase "cops are pigs" when accompanied by a photo of a police officer. We also had a trending job of a man offering a hundred dollars to his daughter to knock off a MAGA cap from a boy's head, and we were told to ignore that, despite its meeting the criteria for the Violence & Incitement policy.

Of course, we also had the exception given to protect Greta Thunberg from words like "gretarded." A few more examples include the Eric Ciaramella exception, allowing Don Lemon to say, "straight white males are terror threats," on national television, and letting the phrase "straight white males are filth" to stay on the platform.

In my opinion, all these exceptions and rule changes were deliberately made and align perfectly with Facebook's leftist ideologies. Here is the wording of some of the exceptions I just mentioned:

> Raquel Salinas* to Cognizant ESLA Trending
> August 20, 2018
> #trending #ignore #MAGACap
> 232857230754115
>
> Video of man offering his daughter $100 to knock MAGA cap off a boy's head.
> Action: Ignore
>
> [reaction] You, Jessica Martinez†, Patricia Charcon‡ and 99 others
>
> This job fits the criteria for Violence and Incitement, 2.1.c. for mid-severity violence, calls for action for violence. Mid-severity includes "hitting" and knocking off a hat would be mid-severity violence, as it's more than a "slap."

---

* A pseudonym.
† A pseudonym.
‡ A pseudonym.

There is also wording in the policy that disallows "bounties," so the fact that there's money involved makes this video more egregious.[3]

And how about Don Lemon saying white males are "terror threats"? Could any other media personality be so virulently racist and keep their job?

Shawn Browder to Cognizant: North America Team
November 15, 2018

[POLICY EXCEPTION] Don Lemon's Quote About White Males Being Terror Threats

Action to Take on Job: Ignore
Issue or Abuse Type: Hate Speech

Summary: CNN host Don Lemon recently said white men are "the biggest terror threat in this country." This is implying that white men are terrorists and so would typically violate (HS Tier 1-2.1.6 dehumanizing speech). As this is a newsworthy event, FB's content policy team is allowing a narrow exception for this content on the platform.

However, considering recent events in the U.S. there is a likelihood that people might support or agree with Don Lemon's views making their posts violate our hate speech policy, and the exception given to the original content will not be extended to any support or agreement from users.

[reaction] You, Lorenzo Cuevas, Eunice Chacon and 533 others[4]

There are a couple of important key aspects of this last post regarding Don Lemon. First, it's clearly a violation of Facebook's hate speech policy, and this is clearly acknowledged.

Second, they call this a "newsworthy event"? Who decides what is newsworthy and why it is newsworthy? That sounds like the actions of a news organization, not a neutral platform.

Third, they talk about recent events in the United States, so their policy adjusts according to current events.

Last of all, they admit that people might agree with Don Lemon's statement.

Facebook is gauging public opinion and essentially saying it's okay for Don Lemon to be racist because other people might agree with him. Even if they're deleting other comments about racism, Don Lemon's influence is extraordinary, reaching millions of people. The fact that they may be attempting to mitigate the fallout by deleting a bunch of comments is really pointless. Facebook allowed racism and hate to foment on its platform— hate and racism directed toward white people. And I find it hard to believe that they would allow this type of racism toward any other group of people.

Then there was the case of the Israeli sniper I previously mentioned in Chapter 4. Because we know Israel is bad, right? No good press coverage for them!

In April 2018, the same month when Mark Zuckerberg testified in Congress, his underlings were busy making exceptions to help portray Israel in a certain light.

> Susanne Dilma[§] to Cognizant: North America Team
> April 11, 2018
> #Exception #HeadsUp #PerTheClient
>
> TLDR NEWSWORTHY EXCEPTION FOR VIDEO SHOWING AN ISRAELI IDF SNIPER KILLING A PALESTINIAN SOLDIER NEAR THE GAZA BORDER
>
> Action to take on the Job: MAD --Mark as Disturbing
> Issue or Abuse Type: Graphic Violence
> Summary: Newsworthy exception for shares of the video when shared to discuss and raise awareness. The video will be marked as disturbing.[5]

At this point in time, we were deleting videos of people dying. However, this policy was changed at some point later that year, so any videos of murders or accidental deaths were always Marked as Disturbing. This guidance told us to make an exception, per the client, to allow this video. This video shows Israel being the aggressor and a Palestinian soldier being the victim. One must wonder why this killing was given an exception, instead of videos of Israeli citizens being killed during missile attacks.

In addition to exceptions regarding violent videos, Facebook has also allowed more attacks against white males, and whites in general. Initially, when I started at Facebook, the phrase "white trash" was violating, but then

---

§ A pseudonym.

they reversed this decision. Were the Facebook executives sitting around a conference table one day when somebody said, "Hey, guys! I just realized it's not an insult to be called 'white trash'! Release it from the censors to run free in the world!"

On December 5, 2019, we received new guidance on how to treat the phrase "white trash." The post was made by Timothy Westover:[¶]

> Timothy Westover
> 12/5/2019 6:41pm
> #HateSpeech #Clarification #Trash #Filth
>
> This post is to provide new and updated guidance for "trash" and "garbage" as well as to clarify slur + shit. This is an exception made as generally they are used as comparisons to Filth and not comparisons to inanimate objects. The call-out for inanimate objects only being deleted for designated attacks towards woman is still relevant, this is only in conjunction with it to avoid any confusion.
>
> - Men are trash → Delete > Comparison to Filth
> - Women are garbage → Delete > Comparison to Filth
> - Women are pieces of trash → Delete > Designated Dehumanizing
>
> Please use cultural context to establish if a term or phrase that uses trash/garbage but has a different cultural meaning.
>
> - You are white trash → No Action – The phrase "white trash" is used to identify someone as a person who is both financially poor and white. This only establishes a subset, not an attack.
> - This is the most white trash thing I've ever seen → No Action
>
> 1. How would we treat slur + shit?
>
> A. In most instances, shit is being used as a synonym for stuff, so it would be a descriptor of the things the slur does, not a statement on the slur itself. When it is clear that slur + shit is being used as a replacement for slur + stuff action as slur.

[reaction by 192]   Seen by 408[6]

¶ A pseudonym.

The post above says that "white trash" is not violating because it is describing someone who is both financially poor and white, and that this is only a subset of a PC (protected characteristic). That blows my mind. Hate speech really can't get any clearer. There are only two words in the phrase "white trash," *white* and *trash*. You are literally calling a protected characteristic "trash." Would Facebook allow "black trash," "yellow trash," or "brown trash"? Are there no poor black, Asian, or Hispanic people in the world?

I can understand some of the cultural meaning and that it refers to poor white people, but the phrase, we're discussing is "white trash," not "trailer trash." In this phrase it's clearly identifying white people as the target of the attack. Why does Facebook hate speech policy specifically allow attacks on one race? To me, that's about as racist as it gets.

And how did people inside Cognizant respond to these changes? Luckily, I documented those interactions on my hidden video and these are transcripts that I personally transcribed from interactions with coworkers discussing this policy change.

Video 27

Conversation with Tyrell Johnson,** who is black and sat next to me. I told him about the guidance to Ignore "you are white trash":

> **Tyrell:** It's totally an attack.
>
> **Ryan:** I agree.
>
> **Tyrell:** Whatever happened to white trash being under hate crime?
>
> **Ryan:** Yeah up until now, up until this post it was a delete for dehumanizing.
>
> **Tyrell:** Dehumanizing right, and if it didn't fall under dehumanizing, it still fell under bullying.
>
> **Ryan:** Yeah bullying is separate, this just clarifies . . .
>
> **Tyrell:** Well it's a character claim under bullying, at least it was anyways.

** A pseudonym.

Ryan: It still is in bullying, especially with a face match, but even in the post it says it's still identifying someone as a person who is both financially poor and white, so it's still identifying based on PC, so that's my issue with it.

Tyrell: You don't call a black person white trash.

Ryan: That doesn't happen right?

Tyrell: No, I've never been called that a day in my life.

Ryan: So I'm gonna ask about that one. It doesn't surprise me.[7]

My black coworker Tyrell agreed that calling someone "white trash" is an attack, and he was just as shocked as I about Facebook's reversing its rules and making "white trash" nonviolating.

VIDEO 36

I requested help on a job from Alfredo, who was a member of the Phoenix policy team. He was doing floor support and helping representatives perform their jobs if they were unsure how to action the job. The job was an image of Frederick Hopkins from South Carolina, who killed one cop and injured seven others:

Ryan: So this is the man who killed a police officer and wounded seven other cops. Yet the media wants you to know that he's a veteran from the war, and then white trash privilege at its finest. So, we have that new callout, just from this morning, er, from Tyler, so it says, You are white trash, no action. The phrase white trash is used to identify someone as a person who's both financially poor and white, this should only establish a subset and not an attack. So here, you have an image of a white guy, it says white trash privilege at its finest, so black unity, black love, so I think based on, what do you think Alfredo on this, cause this guidance is basically saying?

Alfredo: Yeah, but only because his definition is someone that's a person who's both financially poor and white.

Ryan: So is this targeting—

Alfredo: This one is just targeting him for being white, and a trash human being at this point, not merely white trash.

Ryan: But here they're saying, the definition of white trash means a poor person and white.

Alfredo: That's why I'm just like, [unintelligible] because it doesn't meet the definition of poor and white. It meets the white part but, financially poor, I wouldn't say so.

Ryan: Probably still delete this then?

Alfredo: Yeah, for attacking him on his PC [Protected Characteristic].

Ryan: Yeah, I have a lot of questions on that.

Alfredo: Yeah this one's, I looked over this one, I was like, this is gonna, it's a little weird, because it's really just on I guess cultural context, and how you're reading it at that point. But this one, I'm just reading it as just dehumanizing attack, rather than calling him a white trash person.

Ryan: Because for me like when you say white trash—

Alfredo: And they're talking about privilege also, but it's more of an attack like, on him being white than him being white trash.

Ryan: Okay.

Alfredo: I would just go with dehumanizing, just seems the more logical way to go with it.

Ryan: Because for me when people say white trash, I think like, oh like trailer park trash basically.

Alfredo: I agree with that. Just with that context, I feel like it's attacking his PC, more than the definition they gave here. I don't agree with that [pointing to policy page], this probably should not, I honestly want them to get rid of that.

Ryan: Can you imagine if it said black, like we'd still delete like black trash obviously or any other race, right?

Alfredo: Uh-huh.

Ryan: But they're just giving exception for white trash.

Alfredo: For white trash, only because it has a cultural meaning, for meaning something else, which I understand but, at the same time, it shouldn't be there. Yeah but that one I feel it's just more the dehumanizing side, targeting him and calling him a white trash person per the definition.

Ryan: Okay, thanks.[8]

This conversation with Alfredo was about a job I was working on that attacked someone and said they had "white trash privilege at its finest." We interpreted this to violate the hate speech policy because it talked about privilege, so it was attacking the race in general. But even Alfredo, who was on the policy team, disagreed with how they changed the policy to allow "white trash." He said, "For white trash, only because it has a cultural meaning for something else, which I understand, but, at the same time, it shouldn't be there."

Alfredo also pointed at the policy page and said, "I don't agree with that, this probably should not, I honestly want them to get rid of that." So let's summarize Facebook's exceptions regarding race.

"Straight white males are filth for not supporting LGBT" is allowed.

"White males are terror threats" is allowed if you're CNN's Don Lemon.

"White trash" no longer violates the hate speech policy.

And in December 2020, Facebook confirmed this publicly, that they would be deprioritizing hate speech against whites.[9] It's clear from the examples I've presented that they had been doing this long before the public announcement.

Child Abuse Policy Exception

At some point during my time at Facebook, they modified the child abuse policy and made it stricter. Previously we would MAD (mark as disturbing) child abuse. But then a change was made and we were to delete any instances of child abuse, which consists of multiple slapping/hitting of a child.

Senator Fraser Anning is a right-wing politician from Australia. He is notorious for speaking out against Muslim immigration to Australia and even made some comments about the New Zealand Christchurch shooting being a consequence of mass Muslim immigration. He tweeted, "Does anyone still dispute the link between Muslim immigration and violence?"[10]

As a direct response to this statement, Senator Anning had an egg cracked on the back of his head by a teenager. He responded by turning around and slapping/hitting the kid several times.[11]

This action by Senator Anning meets the definition of child abuse, and the whole video should have been deleted. However, Facebook made an exception and allowed this video of child abuse, in clear violation of their own policy.

9_12_2019 AM Video 7: 4mins, 19 secs

Bangladesh-OS Workflow and policy support

Tesani Haokee[††]
March 22 2019

Exception for the viral egging video and clarifications on captions

TLDR CHILD ABUSE: -KID WHO EGG'D SEN FRASER ANNING. Action to take

on job: ignore. Issue or abuse type: Graphic Violence. Summary: Videos of the minor who stuck the senator in the head with the egg would normally qualify for child abuse (senator punched a minor) and torture of humans (guards pinned down the minor). HOWEVER, as this video has newsworthy value (it is part of the current debate and being shared via news agencies) we will allow this video without an interstitial.

- **Egging video + neutral caption/no caption = ignore
- **Egging video + positive reinforcement of the boy's actions(egging) = ignore

---

[††] A pseudonym.

e.g., caption Look at this teen who egged the stupid senator [smiling emoji] or salute to this legend[laughing emoji]

- **Egging video + any positive reinforcement of the boy getting beaten up = delete **

e.g., caption: Good job Fraser Anning, by teaching the kid an instant lesson, with a good slap Senator still got some juice in him.

P.S. Feel free to reach out with any questions/clarifications

[liked by 55]   Seen by 595[12]

What are the possible reasons Facebook allowed this video?

In my opinion, the first reason is it shows something bad happening to a public figure Facebook doesn't like, specifically, a right-wing politician who has criticized immigration policies. I'd argue that this gives the public tacit permission from Facebook that such actions are permissible, if not praiseworthy.

Second, it shows that same politician overreacting to the egging by beating up the teenager. Do you see how it's creating a narrative with several different parts? The politician is a bad person because of his views, and that's reinforced by how he treats the teenager who eggs him.

Let's conduct a thought experiment and place Don Lemon in a similar situation. After his statement that white men are a terrorist threat, imagine that he was egged by some teenager, and he responded by punching the teenager. I think we can agree the media would portray Don Lemon as a hero for standing up to this assault.

Moving on to another group, let's see how transgender people are given special rights by Facebook.

This guidance is for a chart that shows all the symbols for different genders and is referring to all other genders besides male and female as mental disorders. They changed the guidance on this job multiple times. For a while it was an ignore, but around mid-July of 2018, the guidance was to delete. Technically, though, there's no PDITI (person depicted in the image) or stick figure that would represent the Protected Characteristic (PC). All we have is the symbols. Therefore, they made an exception against the policy as they themselves acknowledge:

Transgender Policy Exceptions

Video 7

Cindy Fredrickson[‡‡]
July 13, 2018

#PolicyException #Only2Genders #VisualHate

POLICY GUIDANCE FOR IMAGE REFERENCING ALTERNATE
GENDERS AS MENTAL DISORDERS

FB has made a policy exception for actioning jobs that contain this image
depicting male and female gender symbols as "genders" and all other symbols
as "mental disorder."

This image had been an ignore for not containing an actual PDITI or human
like figure. Based on FB's exception jobs containing this image will now **be
Delete > Hate Speech > T2 Statement of Inferiority**

[Liked by] you and 660 others   1 Share   Seen by 668[13]

Video 7
7 minutes 10 secs

Suzannah Fischer shared a link
October 22, 2018

#HeadsUp #Trump #Transgender

Trump to Exclude the Transgender Community from Federal Civil Rights

Summary
The Trump administration is potentially rolling out a new legal definition
under the Department of Health and Human Services to strip federal recognition of the gender identity of some 1.4 million Americans—and require
genetic testing in some cases to match a person's gender with the sex they

---

‡‡ A pseudonym.

were assigned at birth. In other words, the government would not recognize a person's gender other than the one based on their genitalia when they're born.

Potential Violations

- Hate Speech attack the transgender community as a whole and members of the transgender community
- Harassment against individuals and groups of people
- Bullying against private individuals

Related Articles

https://www.nytimes.com/2018/10/21/us/politics/transgender-trump-administration-sex-definition.html

https://www.vox.com/policy-and-politics/2018/10/21/18005594/trump-adminstration-transgender-sex-dna-text

[headline] nytimes.com Transgender Could Be Defined Out of Existence Under Trump Administration

[reactions] 525   Seen by 535[14]

Now, I'll be the first to recognize that society's standards can evolve. But how is it that a policy that existed through the Clinton and Obama administrations is suddenly so toxic that is must now be categorized as "hate speech"?

Facebook is clearly demonstrating that they are not a "neutral" platform and are guiding the conversation toward their predetermined choices. At the same time, they claim they are doing no such thing.

Is it because Facebook is so large it can feel comfortable ignoring the law? Why hasn't Mark Zuckerberg been charged with perjury?

CHAPTER FOURTEEN

How Facebook Deals with Abortion and the Pro-Choice Movement

When I first started at Facebook in 2018, we didn't do anything to images of abortion imagery. This was an advantage for the pro-life movement and allowed pro-lifers to get their point across. Thus, more people would see aborted fetuses on Facebook at that point in time.

However, in late 2018 or early 2019, Facebook altered the Graphic Violence policy to include the Mark as Sensitive (MAS) action for aborted fetuses.

Mark As Sensitive (MAS) and Mark As Disturbing (MAD) are both actions you can take within the Graphic Violence policy.

These actions, MAD and MAS, limit the type of content that shows up in the feed and filters out graphic content based on a user's preferences.

"Mark as Sensitive" was also used to mark live birth-giving videos. Originally videos of live births were deleted if they showed female genitalia, but later on Facebook allowed these, but we still had to Mark as Sensitive.

"Mark as Disturbing" was used for videos of car accidents, live executions, beheadings, or even wounds that showed visible innards. Videos of these types of situations were deleted, while images were allowed but marked with MAD.*

---

* Facebook's usage of the word "imagery" within the "Graphic Violence" policy referred to both photos and videos, whereas "image" just referred to a photo. This is one of many examples of how the policy language was very complex and nuanced.

Going back to abortion, many of the policies didn't include fetuses or consider fetuses as "private individuals." The "private individuals" language is seen more under the Bullying policy, for example when calling for someone's death. Here is Facebook's guidance explaining that abortion is not considered a violent death:

Henry Marner[†] to Cognizant CO Trending
July 7, 2018
#fetus #abortions #ignore

Hey Team, just a reminder we learned from the FB team that abortions are not considered "violent death" so they would not be part of GV 7 or 8. Abortions are not covered under Cruel & Insensitive for premature death. Following examples would be ignore.
2439216096092184 See more

[reaction] You and 497 others[1]

We also have a second screenshot telling us that a joke about fetus soup is allowed.

Ofelia Montenegro[‡] to Cognizant CO Trending
October 9, 2018
#Trending #FetusSoup #IgnoreBenign #Fetus

559443577627154
This job has been trending today, and depicts someone holding a bowl of soup with a supposed "fetus" in it. Even though we have mocking context, we do not take fetus's for C & I. And since this doesn't have any abortion or abandonment context nor visible innards/dismemberment, which is needed for a MAS, we should ignore-benign this content. Also upon research, this is part of an artist's art piece and is fake. The artist's name is Zhu Yu, you will be able to find the photos referenced in this job by searching his name + "eating fetus art piece"

[reaction] You and 562 others[2]

---

† A pseudonym.
‡ A pseudonym.

This joke about fetus soup is not covered under Cruel and Insensitive (C & I) because there is no dismemberment (but how do you make soup without cutting things up?), and supposedly it's from a work of art. This is the same policy that would cover, for example, jokes about victims of the Boston Marathon bombing, but it apparently doesn't cover fetuses. It's similar to the Graphic Violence policy and the Cruel & Insensitive Policy (C & I) but also has separate sections depending on whether you need to delete the content entirely or simply flag it/mark it. Within Cruel & Insensitive, we have Tier 1—Delete, Tier 2—Mark as Cruel (MAC), and Tier 3—Mark as Insensitive.

Tier 1 is when you have a real photo of someone who suffered a tragedy, violent death, etc. For example, if you were to Photoshop survivors of the Boston Marathon into a meme and it showed them visibly experiencing their injury, then that would be deleted.

Tier 2 of C & I is when you have a photo of them but they're not visibly experiencing the tragedy. For example, just a photo of Robin Williams next to a noose would fall in this category and would be actioned Mark as Cruel (MAC). It would stay up on the platform but be limited in some fashion with its distribution.

Tier 3 is just mocking the concept of a violent tragedy or event. This could be written words without any imagery that's making fun of someone. Cruel & Insensitive says under Tier 1 that "We remove content that depicts real people (see exceptions here) who are visibly experiencing and being laughed at, made fun of, or further degraded for any of the following:

1. premature death
2. serious physical injury (including mutilation)
3. physical violence (including domestic violence)
4. starvation
5. serious disease
6. disability"[3]

I don't know about you, but abortion seems to fit this policy perfectly. However, it all hinges on how Facebook defines "real people." Apparently, Facebook is saying fetuses aren't real people. That's about the only way the policy makes any logical sense to me. Abortion is premature, there is mutilation, and it is physical violence. Abortion fits multiple criteria for this policy, so why does Facebook not remain faithful to its own standards?

This policy also excludes certain public figures like terrorists, also in addition to public figures before 1900. Are fetuses or stillborn babies considered terrorists? No. Then why are they excluded from this policy?

> Stillborns are also not considered a premature death.

> "#593913434411516
> Action: Ignore > Benign
> Reason: Interim to not treat stillborn as premature death"[4]

Therefore, we consider "stillborn" as a baby since it was birthed, but would you consider it a premature death? Another piece of guidance given to content moderators regarding abortion says the following:

> 3. Is advocating on killing babies/fetuses in an abortion context considered
> V & I [violence and incitement]? Or ignore since they're referencing fetuses?

> Advocating for killing babies/fetuses in an abortion context should be ignored.

> #312748932982911- Ignore Benign, abortion context established in parent object.

> [reaction] You and 521 others   Seen by 650[5]

Seth Gruber is a pro-life activist whose video was deleted. I reviewed a user post about this and raised it up to policy manager Shawn Browder as a possible public relations (PR) fire. This was his reply:

> Thanks Ryan,

> I've raised a task with FB (T62476964) to investigate and potentially restore
> the content if needed. Not having a JID [job id] may hinder FB's ability to
> restore the content, however, as going to the individual's personal profile to
> review deleted content may constitute a violation of data user privacy rules.

> Regards,
> Shawn Browder
> Policy & Training Manager
> Technical Ops—Digital Operations

Cognizant"[6]

Contrast Facebook's concern with privacy in this instance with how Facebook scoured the Internet to delete anything attacking Greta Thunberg. If they found posts using the AI, how is that a violation of privacy different from a manual inspection of a user profile? One might think having an AI search thousands of posts is less invasive than one employee viewing a user's private message on Facebook, but the AI would actually have a greater impact on the platform than a single employee breaking Facebook's so-called "privacy agreement." Facebook moved mountains to protect Greta but wouldn't lift a finger to restore a single post by a prominent pro-lifer.

Facebook's inaction to restore content is itself an action.

Granted, I didn't have the job ID of the exact content in question that was deleted, just a secondhand account of a video being deleted. And Mr. Browder did send a task to FB, so there does seem to be a process for correcting such mistakes.[7]

In an article, on liveaction.com,[8] Facebook is quoted as stating that the content was deleted under the spam policy and that he had been posting repeatedly. Here is the transcript from a video I filmed of Facebook's spam policy:

> The type of content that does **not**[§] violate our policies and is therefore allowed on our platform.

> 3. "Activist Spam:" Repetitive page posts for a political/social cause[9]

Another great example of Facebook's bias is when they gave an exception to Hollywood celebs who were advocating politically against an abortion ban in Alabama:

> Bill Stelter[¶] to Cognizant: North America Team
> May 17 at 9:46am
> #HateSpeech #Exclusion #Exception
> #285763939042973

> We are seeing content trending around the recent anti-abortion law passed in Alabama. This image in particular has been raised to the client, as it meets our Hate Speech policy for Political Exclusion. However, given the newsworthy

---

§ Emphasis added.
¶ A pseudonym.

nature of the content, client has given us direction to IGNORE this image. As always, please be aware of further violations in captions and comments.

[reaction] You, Shawn Browder, Eunice Chacon and 519 others[10]

The phrase they used was "Men shouldn't be making laws about women's bodies." What if there were a similar debate about gun control and a group of conservative celebrities had used the phrase "Women shouldn't be making laws about a man's gun"? Would that have been "newsworthy"?

Alyssa Milano in particular used this phrase many times. But as we can see, Facebook deliberately made an exception to their policy to allow the phrase, despite its clearly violating company policy, by their own admission. To play the devil's advocate, perhaps Facebook felt it was an important debate occurring in society. It was a debate about abortion and could influence other conservative states to adopt similar legislation. But it was also a debate that consisted of rich and famous people in one state telling religious conservatives in another state what to do with their bodies.

I agree that the phrase is somewhat innocuous, but it violates Facebook's policy. Therefore, it should be applied equally, not given exceptions when Hollywood celebrities decide to express themselves. What makes the opinion of some celebrity on a political issue more important than the opinion of an average citizen?

CHAPTER FIFTEEN

Civil War and Boogaloo: What Facebook "Puts in Place" to Prevent Harm

In December 2019, because of the Impeachment Proceedings against President Trump, one of the trends we had was people discussing whether this might trigger a civil war. The word *boogaloo* was often used, sometimes in conjunction with the mention of civil war. The idea was that if Trump were impeached, there would be a civil war from those on the right. To provide a little background, the word *boogaloo* used to refer to a dancing style and was popular in the 1960s:

> Sometimes stylized as *bugalu*, boogaloo is both a dance and musical genre. It's most popularly a mixture of Latin styles, such as mambo, cha cha and pachanga, with doo-wop and soul. And it's upbeat and easy to dance to—a freeform dance where bodies jerk in time to the music. Elbows and arms are thrown to the sides or over the shoulders, and dancers are free to add as much fancy footwork as they can."[1]

Now, however, *boogaloo* is used to refer to a civil war. The NPR definition refers to it as a racist term, but the Urban Dictionary definition doesn't mention that, and I never saw any evidence of racial undertones with the word. Here is the first entry from Urban Dictionary:

A get together of people ready to go to war against "liberal lefties." Ususally meme'd as a joke by right wing people or those actively trying to keep their firearms when Democratic politicians go for their guns as a time when they have to defend their right to bare arms by proclaiming "Try me."

You hear how the governor of Virginia was trying to take everyones guns?!! *sigh* Time to start the "Boogaloo" again.[2]

Another entry from Urban Dictionary:

When the United States citizens rise up against their tyrannical government to ensure **liberty**, and freedom for all.

The boogaloo *is going to* happen *you* ready?[3]

By looking at a few Instagram pages, you can see what Boogaloo memes look like. And some of the people advocating for Boogaloo refer to themselves as Boogaloo Bois. Here is a transcript from my conversations with my coworker Frank* and my manager Shawn.

Shawn Browder was the policy manager for the Phoenix site and oversaw policy decisions for over a thousand individuals. He was a Cognizant employee and had daily video conference sessions with the client, Facebook.

Video 8   2019/12/26 10:05:32

Shawn: This is kind of gonna boil over. [talking about calls for civil war regarding impeachment] And could lead to real world harm, if there wasn't **something we put in place**, regarding this type of content. Maybe an exception under policy violence & incitement, coordinating harm . . . but just keep sending them [job ids], so that they can start seeing all the stuff that's happening. I'm gonna take a look at the jobs you sent and if it's the civil war stuff, I'm gonna send it straight to Facebook.

Ryan: Is the boogaloo serious or are they joking?

Shawn: Some people are joking, some people are serious. That's the danger of leaving it on the platform is, where's that line.

Shawn: Definitely the civil war stuff.

---

* A pseudonym.

2019/12/26 10:06:58

Frank: You can see people hyping each other up.
2019/12/26 10:07:05

> **Shawn:** That's exactly the type of stuff that Facebook wants to see, like had
> they seen that type of stuff, like leading to the 2016 election, like they would
> have definitely like put some things in place, to prevent it.[4]

Boogaloo is a term that loosely means the same thing, a civil war or revolution. There are militias that use this word. Also, many Instagram users use this term to connect and organize gear and equipment. It's hard to tell how serious some of the users are. There are many memes related to Boogaloo.

For example, there's an Instagram user named @weikleforsenate who is presumably running for Senate in South Carolina.[5] He posts a lot of funny memes, but at one point he also posted instructions from public source material (army training manuals) with instructions on how to create DIY explosives. As you can see, it's a weird mixture because on the one hand you have very funny memes. But Weikle is also posting information that could be used in very bad ways.

I like the fact that memes can be used to educate people about current issues. But I personally don't feel comfortable sharing instructions for weapons that could be used for harm.

I wonder why Facebook was not as concerned about calls for Civil War from the left after Trump threatened to replace Ruth Bader Ginsburg on the Supreme Court. They were very concerned about Republicans threatening Civil War over Trump's impeachment, and they were very concerned about the Gun Rally in Virginia on January 20, 2020:

> Shawn Browder to Cognizant CO Trending
> January 20 at 9:15 AM
> #trending #gunrally #virginia
>
> Summary: A gun rally is currently underway in Virginia. Governor Ralph
> Northam declared a state of emergency starging 1/17/2020 and ending
> 1/21/2020. Current news states that the majority of the protesters have been
> peaceful, but there are reports of The Base (a neo-Nazi terrorist group) present
> at the rally.

https://www.washingtonpost.com/dc-md-va/2020/01/20/virginia-gun-rally-updates

[reaction] You and 296 others[6]

If Facebook is flagging "boogaloo" content to prevent real-world violence, then I understand. But if you're limiting and restricting "boogaloo" while letting Antifa run rampant on your platform, then there's not much fair and balanced progress being made.

Here is an example of how a small business using "boogaloo" for marketing purposes was completely deleted off social media. They were even careful to avoid "boogaloo" memes that insinuated violence. Justin Nazaroff is the manager of Fenix Ammo, and he emailed me about what happened to him when he starting using "boogaloo" memes as a way to market to the Second-Amendment crowd:

Thanks for replying to me on Twitter! As I said - I really enjoyed the podcast. My guys here at the shop listen to Tim's podcasts almost every day on the production floor and your insight into the world of tech censorship was a really great addition.

It was particularly interesting because we have been banned on both Facebook and Instagram. As I mentioned, we had been using the "boogaloo," "big igloo," "Hawaiian Shirt Party," and other such terms/memes dating back to at least August of 2019. At that time we had around 7,500 Facebook followers and approximately 30,000 Instagram followers. Not bad for a small business. I personally posted all of the content myself and I was fairly proud of the fact that we had built a little following in a totally organic way.

Most of what we posted consisted of product photos, manufacturing videos, and reposted memes that I found on other pages. We definitely went "all in" on the Boogaloo type memes although I was careful to avoid posting things that I felt directly insinuated violence toward any particular group. The closest I came was a post in June that was a reply to a woman who was suggesting that the George Floyd rioters come out to the suburbs here in Michigan (where we are located).

In any case - we never received any warnings from Facebook or Instagram regarding our content. We never had a post removed for violating their policies. Ever.

A "journalist" for an anti-gun website, The Trace, decided to write an article about the Boogaloo memes and I was dumb enough to talk to him.

Once he published the article our Instagram page was deleted the very next day, without warning. My personal Facebook account was locked out because I was the administrator. I reactivated an old Facebook account I had so that I could continue to manage the business Facebook page. Two days later, it was deleted as well. They also locked my personal account, again, and also locked the accounts of my brother and two of our other employees who were listed as administrators for the FB page. They got their accounts reinstated within a day. My account, however, is STILL locked, and continues to say that it's "pending a review."

We had the lawyer for Steven Crowder send a couple letters to Facebook asking them to reinstate our accounts, but we did not get a response. We honestly have no real idea what caused our pages to be deleted because we never even received a notification - they were just gone.

Also - they seem to be tracking the IP address on my devices, because I tried to create new personal accounts and it will not let me do so. I did create a new account for Fenix a day after we got banned using a different laptop, but that account also got deleted less than three hours later.

I'd really like to know what went on in the background with boogaloo type content and when they started to crack down on people like us who were using it in a cheeky, fun way. We have no affiliations with any "white supremacy" groups, obviously (should go without saying, but), and in fact I did a pretty good job of attacking all sides. We regularly criticized police, and supported numerous black gun owner groups and still do to this day.

In any case, thanks for listening to my little story, and let me know what you think! If you have any ideas on how me might get our accounts back that would be great although, I am sure that's not possible.

Stay safe and continue doing what you're doing. It's essential during these times. We like to think of ourselves as "single issue voters" when it comes to the Second Amendment, but, I see the First and the Second Amendments as being inextricably linked. You can't have one without the other. Thank you for doing your part.[7]

As of April 2021, the Facebook and Instagram accounts for Fenix Ammo are still locked. One wonders how long a platform can continue to ignore their customers until people find or create new platforms.

Facebook's Policies regarding Elections and Violence

Facebook already has a policy in place that prevents "violence due to the outcome of an election," within the Violence and Incitement policy. The policy states: "High or mid-severity violence due to voting/elections outcome: Delete (credibility not required)."[8]

It seems that a lot of protests after Trump was elected involved violence, but I never received guidance to delete any Antifa protest or similar violent protest during my two years at Cognizant. This policy about election violence requires high or midseverity violence. For example, if I said, "If Trump wins I'm going to beat up my neighbor Fred," then that would violate. Or if you said, "If Biden wins I'm going to kill someone," then this would also violate.

However, simply calling for a civil war, such as "If Biden wins, we need a civil war" was permissible. There was nothing in the policy at the time that would tell us to delete that phrase.

That phrase isn't specific enough to delete. Another relevant policy is the Coordinating Harm policy, which prohibits encouraging or committing vandalism or theft, among other things.

The policy prohibits them in Section 2, Harm Against Property:

1. Coordinating (statements of intent, calls to action, representing, supporting or advocacy) OR depicting, admitting to or promoting the following acts committed by you or your associates
   1. Vandalism
2. Hacking when the intent is to hijack a domain, corrupt or disrupt cyber systems, seek ransoms, or gain unauthorized access to data systems.
3. Coordinating, (statements of intent, calls to action, representing, supporting or advocacy) OR depicting, admitting to theft when committed by you or your associates, as well as speaking positively about theft when committed by a stranger.[9]

As you can see, the section about theft is broader, because you cannot speak positively about theft even when committed by a stranger. However, when talking about vandalism, it's only violating if you mention that it was committed by you or your associates:

"I love all the vandalism Antifa did to Baltimore. That racist city deserves it"
= allowed.

"I love that my friends stole so much shit from those stores in Atlanta over the weekend" = deleted.

How Facebook Treated a Death Slogan against President Khamenei of Iran

How does Facebook treat death threats against Iran's Supreme Leader? Oddly enough, for about two months, Facebook allowed the phrase "Death to Khamenei" in Iran:

> Shawn Browder shared a post
> January 6 [2020] at 6:01am
> Sharing for visibility
>
> Newsworthy exception lifted
>
> Hi all, since demonstrations are over, Facebook has confirmed that we are lifting this exception. In line with status quo policy, threats to heads of state should be removed/escalated (i.e., Death to Khamenei = Violating). Please let us know if you have any questions.[10]

An earlier post in November gave the initial guidance:

> Thad Franklin† to Cognizant: North America Team
> November 20, 2019
> #Exception #Khamenei #Slogan T57853052
>
> TLDR- EXCEPTION TO ALLOW DEATH TO KHAMENEI SLOGAN UNTIL FURTHER
> NOTICE
> Action to take on Job: No Action
> Issue or Abuse Type: Credible Violence
> Specific Violation: High Severity Violence
> Summary: Demonstrations erupted in Iran after the government announced that fuel prices would increase by 300%. The demonstrations have spread across Tehran, Ahraz, and Baluchistan. Though the demonstrations are relatively peaceful, according to media reports, several protesters have been killed and injured. "Death to Khamenei" is one of the main forms of political

---

† A pseudonym.

expression and a slogan demonstrators use to show anger about the fuel price increase.

Under our Violence & Incitement policy, we delete content aimed at public figures that are statements of intent to commit high-severity violence or calls for action of high-severity violence. "Death to Khamenei" meets this definition and would ordinarily be deleted under our policy. However we would like to grant an exception to allow the slogan's usage until further notice.

[reaction] You and 306 others   Seen by 788[11]

This shows Facebook adapting their policy. In this instance, Facebook is supporting free speech because of mass protests in Iran. However, it was just a temporary exception. The phrase "Death to Khamenei" was temporarily allowed for nearly two months from November 20, 2019, to January 6, 2020.

What is striking about this is that it shows Facebook's influence. Facebook can unilaterally decide if a mass protest goes viral on social media. Technically, the phrase "Death to Khamenei" violates their policy, so they could have chosen to not allow mass protests to be broadcast on their platform.

The question remains, who at Facebook decides when a protest warrants an exception?

Although Americans may like the fact that the phrase "Death to Khamenei" is allowed, the bigger question has to do with Facebook's influence. If they can allow mass protests, they can just as easily disallow trends or slogans.

If Facebook is willing to intervene in Iran, it makes you wonder what actions they're taking in America, whether it's against "Boogaloo," civil war, or other notorious individuals who are involved in violent acts of self-defense.

Kyle Rittenhouse and Facebook's Treatment of Him

On August 25, 2020, Kyle Rittenhouse was at a riot in Kenosha, Wisconsin, and shot three rioters in self-defense.[12]

This was one of the biggest news stories of the year, and the media didn't let this opportunity go to waste. They smeared Kyle Rittenhouse as a white supremacist lone wolf who was intent on murdering protesters. *The Federalist* said:

The information in the "situation" section that NBC cited as talking points, the DHS officials noted, is actually background information based on news reports of the incident. It explains that Rittenhouse, 17, is being charged with first-degree murder, and "his lawyers are arguing self-defense." It also states that Rittenhouse "took his rifle to the scene of the rioting to help defend small business owners" and that "he was seen being chased and attacked by rioters before allegedly shooting three of them."

The situational background outlines how media and Democrats "are trying to craft the narrative of a police-obsessed lone-wolf" and a "white supremacist" using "zero evidence." This section of the document also suggests that "subsequent video has emerged reportedly showing that there were 'multiple gunmen' involved, which would lend more credence to the self-defense claims."[13]

Instead of waiting for the dust to settle, or sticking to the background information from the Department of Homeland Security, Facebook aided these false narratives and attacks against Rittenhouse, who is a seventeen-year-old minor.

I created a PowerPoint about Kyle Rittenhouse and my theory on how Facebook classified Rittenhouse as a mass murderer. This is the only explanation for why any positive mention of Rittenhouse was instantly deleted by Facebook, and why fundraisers or other support was strictly prohibited on the social media platform.

What is the policy that pertains to Rittenhouse? Is Facebook classifying him as a mass murderer? The Dangerous Individuals and Organizations policy is what Facebook uses to make decisions about terrorist groups, racial supremacists, hate groups, cartels, and mass murderers.

The policy states:

1. We do not allow the following people (living or deceased) or groups to maintain a presence (e.g., have an account, Page, Group) on our platform:

    1. Terrorist organizations and terrorists
    2. Hate organizations and their leaders and prominent members
    3. People notable for physically attacking people based on a protected characteristic
    4. People who have committed or attempted mass murder
    5. People who have committed multiple murder
    6. Criminal organizations and their leaders and prominent members

2. We do not allow symbols that represent any of the above organizations or individuals to be shared on our platform without context that condemns or neutrally discusses the content.

3. We do not allow content that praises any of the above organizations or individuals or any acts committed by the above organizations or individuals.

4. We do not allow coordination of support for any of the above organizations or individuals or any acts committed by the above organizations or individuals.[14]

A new addition to the policy is the phrase in 1.3: "People notable for physically attacking people based on a protected characteristic." This is very vague, and I wasn't aware of a list they had for who would fit this. It seems that Facebook can make its own determination about who is "notable."

My first theory is that Facebook labeled Kyle Rittenhouse under "people who have committed or attempted mass murder." Mass murder is defined in Known Question (KQ) Section G as:

G. What are the definitions of mass murderer and serial murderer?

1. Mass Murderer: 4+ casualties not including perpetrator, during a single incident with no time between murders.

2. Serial Murderer: 2+ casualties from the same perpetrator over multiple incidents or locations.[15]

However, even if you were to operate under the assumption that Rittenhouse was attempting mass murder, he still doesn't fit the definition because it was a single incident and the casualties didn't reach four, as there were only three casualties. But the policy just defines it as "people who have committed or **attempted** mass murder," so it's possible Facebook believed that Rittenhouse "attempted" mass murder. As you can see, there are several ways Facebook could have fit Rittenhouse into this policy.

It wouldn't be difficult, since the policy is vague and has flexibility to accommodate Facebook's own preferences. And everything we saw happen to Rittenhouse matches this policy. Any "coordination of support" from Section 4 above was prohibited. No content was allowed that "praised" Rittenhouse, as mentioned in Section 3. Praise is defined by FB as speaking positively about someone.

The only way you can discuss people who fall into this category is by neutrally discussing the event or condemning the actions or the people involved. In the ignore section of the policy, the following is allowed:

1. Challenging historical events and facts (e.g., "The death toll of Stalin's purges has been grossly exaggerated").
2. Engaging in neutral discussion of a dangerous organization, members, or actions in the context of academic discourse and news reporting.
3. Depicting weaponry outside the context of terrorism, organized crime, or hate.[16]

Another part of the policy says, "We allow video imagery depicting a dangerous entity ONLY if it is shared, depicting, or referenced in the context of condemnation or neutral speech."[17]

Before the trial even started, Rittenhouse got lumped together in the same Facebook policy with terrorists like Osama Bin Laden, and mass murderers like Adam Lanza and the Columbine murderers.

Because Kyle Rittenhouse is now defined as a dangerous individual per FB policy, death threats are now allowed against him, and he is no longer protected under the Bullying policy:

Violence and Incitement Policy (V & I)

Tier 1: High-severity violence- threats of violence that lead to death

1. Content about any target(s), including a public figure, and any of the following:

   a. Statements of intent to commit high-severity violence; OR
   b. Calls for action of high-severity (UNLESS the target is an organization or individual covered in the Dangerous Individuals and Organizations policy, or is described as having carried out violent crimes or sexual offenses, wherein criminal/predator status has been established by media reports, market knowledge of news events, etc.).
      i. including content where no target is specified but a symbol represents the target and/or includes a visual of an armament to represent violence OR

    c. Statements advocating for high-severity violence (UNLESS the target
       is an organization or individual covered in the Dangerous Individuals
       and Organizations policy, or is described as having carried out violent
       crimes or sexual offenses, wherein criminal/predator status has been
       established by media reports, market knowledge of news events, etc.).[18]

So to summarize:

"I am going to kill Kyle Rittenhouse tomorrow at 5pm with my Glock 19" =
deleted

"Someone should kill Kyle Rittenhouse" = allowed

"Let's all get together with butcher knives and chop up Kyle Rittenhouse" =
allowed

"Someone needs to take an AR-15 and shoot Kyle Rittenhouse through the
eyes, then slowly chop off all his limbs and make him suffer an agonizing
death" = allowed

"Someone needs to chop off Kyle Rittenhouse's penis and feed it to the dogs"
= allowed

As you can see, Facebook isn't even trying to accomplish its stated goal of
preventing violence on the platform. It's incredibly easy for Facebook to
subjectively apply its policies when the cultural winds prompt swift action
against scapegoats who have not yet received due process under the law.

    In other words, Facebook isn't trying to stop violence. It's directing vio-
lence toward its preferred targets, just as if it were a soldier on the battlefield
painting an enemy tank with a laser and calling in a drone strike.

CHAPTER SIXTEEN

The Fire Brigade: How Facebook Prevents Public Relations Disasters and Shields Accounts from Deletion

From a business perspective, it makes sense for Facebook to prevent automatic deletion of business customers' accounts. These are paying customers who are in a business relationship with Facebook and pay for marketing, ads, etc. Many companies don't have a website. They only have a Facebook page and use the platform to find new customers, communicate with their followers, and build their business. Entire livelihoods depend on the infrastructure that Facebook has built out for the business world.

But what happens when the needs of commerce conflict with Facebook's liberal beliefs?

What I find shocking are allegations that Facebook is purposely attacking its own business customers by deleting their followers and limiting the distribution of their content.

For example, Joshua Feuerstein is an evangelical pastor who also talks about politics. On his Facebook page, he had 2.7 million followers. He also owned a media agency that spent millions of dollars with Facebook on behalf of his clients. He contacted me in July 2020 and asked me if I'd ever seen his content pop up. I had seen it quite a bit as a content moderator. Later in the year, his account was deleted, then restored.

However, when he would post new videos, the amount of engagement was dramatically reduced. Individuals like Joshua who are conservative activists often spend thousands of dollars on marketing, all paid to Facebook. And in return, Facebook restricts their content or deletes their whole page.

Another group is Great State Alliance, which is a registered nonprofit in Arizona. I spoke with the director, Jeremy Wood, who said that all their followers are organic, but they lose followers every day. The attrition of followers didn't make sense, and so Mr. Wood decided to create a separate website for their organization. They still have a presence on Facebook with over four thousand followers; however, Facebook has removed three other groups belonging to their organization.

The stories of censorship continue, and Facebook is alleged to censor large conservative organizations on the platform, often sabotaging and restricting usage for its business customers. How can Facebook get away with this, and does it hurt their bottom line?

Apparently, Facebook is wealthy enough that they don't need the additional business revenue from conservative activist pages. They have shown a complete disregard for basic fair business practices and seem to suffer extraordinarily little because of this irresponsible behavior. What I did discover, however, is that Facebook does have a mechanism in place to prevent accidental deletion of business partner accounts. Here are the different shields and tags that can be applied to accounts on Instagram or Facebook:

9_27_2019 Footage

Video 2, 6 mins 45 secs

This content is protected by a Hi Pri IP XCheck. Pass the content to the appropriate FTE queue for additional review.

This content is shielded by Fire Brigade and therefore cannot be actioned. If you believe this content violates policy, you can escalate to fire brigade for review.

This content is protected by a mid-pri XCheck tag, please follow the appropriate XCheck guidance.

This content is XCheck tagged with a MediaOps High-Pri Protection Level and therefore cannot be actioned. If you believe this content violates policy, you can escalate for review.

> This content is protected by a media ops absolute protection XCheck tag and therefore cannot be actioned. If you believe this content violates policy, you can escalate for review.
>
> This content belongs to a user or page identified by Media Operations as a business partner and therefore cannot be actioned. If you believe this content violates policy, you can escalate to Fire Brigade for review. If you're unable to escalate the report in CRT, please create a task with all necessary info and assign to the Media Operations on-call found in the on-call tool.
>
> This object has absolute protection against disabling actions"[1]

This explains how "business partners" can receive special protections that prevent certain associates from actioning on their content, if they are identified by Media Operations as such. We have Hi Pri, which means high priority, and we also have something called Fire Brigade, which I assume refers to possible PR fires if the content were to be deleted. Since we as content moderators cannot action the content, we are advised to escalate it if we believe it violates policy and it spells out the process to follow in these situations.

So clearly there are safeguards in place to protect Facebook's valuable business partners, but it's clear from anecdotal examples that Facebook does not protect its business relationships with many conservative activists.[2] I did find a few specific examples of how the "shield protection notification" appeared on certain accounts. For example, I documented that a left-leaning news site named *Now This* reposted a CNN interview with Mike Pence where he allegedly refuses to call a KKK wizard "deplorable." In the top-left corner of the screen an icon appears that says, "shield protection notification," and when you click on it, it shows the types of protections afforded by the Fire Brigade, as listed above. In another example I documented, a picture posted of a Coca-Cola-themed birthday cake also has the same "shield protection notification."

One of the other labels I saw is "Xcheck tagged," which is also listed above as a form of account protection. This one I didn't see as often, but I documented that it was applied to Demi Lovato's account.[3] Shaun King, a prominent leftist activist, has the "Xcheck tag," as well. I also saw that an advertisement for a new *Batwoman* television series from the CW Network had the "Xcheck tagged" label on their account. Another individual with this same tag is Joro Olumofin, who is a Nigerian influencer known as the "love doctor." In the post I filmed, Joro discusses making love to a woman on her period and someone comments below suggesting that is rape. Joro has a history of making inappropriate comments toward women and at one point apologized for his outburst.[4] He currently has nearly nine hundred thousand followers on Instagram.

Another system in place that I documented labels Facebook accounts "verified" or "shielded." While reviewing content, I could see whether someone's account was "verified," "shielded," or both. I filmed more than forty accounts that had these labels and made an Excel spreadsheet to look for trends. Of these forty accounts, I didn't see any consistent pattern. The only anomaly I discovered was that Dinesh D'Souza's Instagram account with 543,000 followers was neither verified nor shielded.[5] Other accounts like Prager University had both protections, and so did Ben Shapiro's account. However, the prominent Black Lives Matter activist Shaun King was not "shielded" but did have the "Xcheck tag."

I didn't come across the "Xcheck tag" very often, so it was hard to find a clear pattern, and it's also unclear what "verified" and "shielded" actually do to the account and how this is different from the "shield protection notification" and the Fire Brigade.

In conclusion, there is still a lot we don't know about Facebook's system for protecting business accounts. The types of protections I documented give us more insight into the internal mechanisms they use to keep content from being removed, yet it's not clear whether this particular system is being used to oppress conservative voices. Some content has "absolute protection against disabling actions," but how many conservatives didn't receive the benefit of this safeguard? At the very least, we know that a system exists to prevent deletion of a user's material and protect certain accounts. A more thorough investigation would be necessary to definitively say that this Facebook tool is being misused.

CHAPTER SEVENTEEN

The Future of Censorship: Section 230 and the Communication Decency Act

The year was 1996, and legislation was just enacted as part of Title V of the Telecommunications Act of 1996. Part of the original intent of that legislation was to protect children from viewing "obscene or indecent" material published on the Internet. Well, let's just say that failed miserably and has raised a host of other problems not foreseen at the time.

There are several possible solutions to this quagmire involving Section 230 and how to deal with Facebook censorship and monopolistic practices. Section 230 is part of a law from 1996 that allows social media companies to remove objectionable content from their platforms, without being liable for those censorship actions. The law was designed as a way to let the Internet grow without libel or defamation lawsuits hampering online forums or other websites. It was also designed to protect children on the Internet from viewing obscene or indecent material.[1]

In the last twenty years we've seen the Internet grow tremendously, and a few big players like Google, Amazon, Apple, Twitter, and Facebook now control large swaths of the Internet and abuse their position of power, using Section 230 to protect their brand and censor political speech. Both sides of the political aisle disagree about how it should be fixed. It's almost like the Goldilocks Porridge problem, because the left thinks Section 230 isn't doing enough to censor content (porridge is too cold), while the right "claims it allows for too much censorship" (porridge is too hot).[2] Some believe Section 230 should be reformed, others believe it should removed completely, and still others just think it should be enforced as written.

However, apart from the Section 230 debate is the belief that Big Tech has too much power to begin with and they are effectively acting as monopolies. Thus, one approach would be to break up the Big Tech companies through antitrust action. On December 9, forty-eight states filed an antitrust lawsuit against Facebook.[3] Separately, the FTC filed an antitrust lawsuit for many of the same reasons.[4] The lawsuits focus on Facebook's acquisition of Instagram and Whatsapp, among other things. Breaking up Facebook would help dissolve the amount of concentrated power they wield.

Another approach is to reinterpret Section 230. There is a lawsuit that was declined by the Supreme Court that could have fixed Section 230. This lawsuit involves someone named Jason Fyk, who sued Facebook for deleting his page and attempting to sell it to someone else.[5] The docket number for the Jason Fyk case was 20-632, and there are five questions that will be presented:

1. What is the breadth of the Communications Decency Act, and does Facebook get immunity under section 230(C)(2)(A)?
2. What is the difference between Facebook "misbehaving" by itself or acting "as the publisher or speak" for removing content in bad faith?
3. Does the law require Facebook to develop content in part or in whole in order to be ineligible for protections under Section 230?
4. Does (c)(1) of Section 230 protect any decision to edit or remove content?
5. If the ICSP (Interactive Computer Service Provider) develops, even in part, the publisher's content with an anticompetitive animus, is the ICSP acting as a "Good Samaritan" eligible for CDA immunity?[6]

These are all great questions and need to be answered. Jason Fyk's argument is that the courts in California have misinterpreted Section 230, granting Facebook additional protections. The Communications Decency Act (CDA) has never been interpreted by the Supreme Court, so assuming they interpret it correctly, this could resolve some of the issues we have with Facebook censorship.[7]

Fyk's approach to clarify Section 230 is the traditional legal route. Jason Fyk filed a lawsuit against Facebook in 2018 alleging anticompetitive behavior. The California courts determined 230(c)(1) does protect Facebook from anticompetitive content restriction, since Fyk sought to treat Facebook as "a publisher." Fyk has also alleged motive matters, but the 9th Circuit Court determined that 230(c)(1) has no measure of motive and dismissed his case. He took the case to the Supreme Court, and it seemed that he had a good chance of being heard when Justice Thomas rendered an opinion in the Enigma Denial of Certiorari that was identical to his argument.

Unfortunately, the Supreme Court did not hear the case, but all was not lost. In an extraordinary turn of events, the *Enigma v. Malwarebytes* decision answered the exact same question that Fyk had asked, whether Section 230 immunizes anticompetitive content restriction decisions. The same court that had previously determined, only four months earlier, that Section 230 immunizes anticompetitive misconduct reversed their position, stating, "the Good Samaritan provision of the Communication Decency Act does not immunize blocking and filtering decisions driven by an anticompetitive animus."[8] Fyk has recently filed a motion for rule 60(b) to vacate his judgment based on the new conflicting precedent enigma created (we should all keep tabs on his progress).

Fyk had been right all along, which leads me to believe his understanding of Section 230 is far more on target than most people recognize. Fyk explained:

> The courts have entirely missed the intelligible principle right up until Enigma. "Good Samaritan" is in quotes for a reason. It is the fundamental principle upon which all regulatory decisions must be based if Section 230 is to be considered Constitutional sufficient. An interactive computer service provider is granted quasi-legislative regulatory authority by the government (230) provided it acts for the good of society. Removing content predicated on their own bottom line or for their own agenda is not the behavior of a "Good Samaritan" and strips the company of any immunity, or at least that's the correct understanding of how it should work.[9]

Another question he raised is the court's misinterpretation of what constitutes an information content provider:

> The definition says, creation or development in part which means an insignificant responsibility, but somehow the courts converted that into a substantially

contribution. There is no other way to explain it other than the court is just wrong. The court's have read extra immunity into the statue where it does not belong. For example, the courts said I was treating them as (a publisher) and granted Facebook 230(c)(1) immunity, it's like . . . hey judge, read the statute. 230(c)(1) says the service provider cannot be treated as (the publisher) not (a publisher). The publisher is the original publisher and a publisher can be in addition to the publisher.[10]

Fyk went on to explain:

It's really simple to understand when you boil Section 230 down to its core elements. First, any action or the omission of action must be in the vein of a "Good Samaritan". Second, If the service provider takes no action at all, to publish, provide, modify or manipulate materials in any way, it cannot be treated as (the publisher) who provided those materials. A lot of people think 230(c)(1) applies to editorial decisions to leave content on the site, nope that's not correct either. A service provider is supposed to be a mere conduit, a passive host of materials. Deciding to leave content is still a publishing action and is subject to publisher liability and finally, if the service provider takes any action, because it literally says "any action" in 230(c)(2) it becomes (a publisher) with one function, to remove offensive materials in good faith acting as a Good Samaritan. That's how Section 230 has to work if the law is harmonious and constitutionally sound.[11]

Jason is still working on fixing this as I write. He is potentially planning to sue the United States itself over Section 230's constitutionality. The courts need to do their jobs and apply Section 230 properly to provide everyone Due Process. He's gone into much greater detail in several op-ed articles and recently proposed a legislative amendment to "FYK-S" Section 230. His proposal is based on resolving the court's misunderstandings. Section 230 is not necessarily broken, but simply too vague to apply uniformly.

To continue to understand possible solutions, let's look at what has been done in the past regarding Facebook censorship.

In May 2018, a civic audit was done by the Covington Law Firm and former Senator Jon Kyl, to investigate whether there was bias against conservatives at Facebook. The results, which came out in August 2019, were inconclusive.[12] I've done a more detailed analysis of this civic audit in the next chapter of this book.

On May 28, 2020, President Trump signed an executive order attempting to curb Big Tech's unchecked power. This order attempted to redefine social media platforms as a "modern public square." It also petitioned the FCC to propose regulations that clarify provisions of Section 230 of the Communications Decency Act.[13]

On December 10, 2020, Rep. Paul Gosar (R-AZ) and Rep. Tulsi Gabbard (D-HI) introduced a bipartisan bill, the Break Up Big Tech Act, which would remove Section 230 protections from Big Tech companies and only allow that censorship if users opt in. Also, any companies selling personal data or otherwise monetizing the platform would be stripped of their protections under Section 230.[14]

Here are two articles I wrote for medium.com which describe and summarize the issues we face against Big Tech and their draconian censorship. From my July 9, 2020, medium.com article:[15]

Excerpts from "A Look at Facebook's Sad and Severe Censorship Saga"

I was a content moderator for Facebook and over the course of nearly two years, I noticed countless examples of bias and political censorship that were given as directives as well as being built into the policy.

The experience gave me no choice but to come forward. Over the course of many months, I documented, with the help of Project Veritas, how Facebook censored conservative viewpoints and promoted leftist ideology.

When I watched Facebook CEO Mark Zuckerberg's April 2018 Capitol Hill testimony, I noticed a stark contrast between his statements dismissing any suggestion that Facebook censored political speech and the long list of exceptions that Facebook issued to content moderators like me. It was an odd dissonance as Zuckerberg testified in front of Congress. I, along with all the other content moderators, were censoring political speech upon his orders.

In 2019, former Arizona senator Jon Kyl completed a so-called civic audit for Facebook in response to complaints from conservatives. Afterwards, I noticed a few things did change. First of all, Facebook began tracking the exceptions to their community standards policy. The exceptions they gave us were now numbered and listed. They also stopped using the phrase "newsworthy exceptions."

This did not mean there was real reform, however. Facebook began using new terminology to allow for them to make decisions about how we actioned content. For example, the word we would often throw around would be "align."

[...]

This bafflement was created by Congress in the Communications Decency Act of 1996, especially its Section 230, which classified Internet Service Providers as different from publishers. Facebook and other social media platforms have taken refuge within this safe harbor protected from libel suits or other sanctions as a legacy of the law.

When Facebook censors content it published as supplied by users, it labels that content hateful, false or inciting violence — which inescapably impugns the content supplier. Everywhere else in the media, a publisher would open themselves up to sanctions, such as a libel suit.

If we take away Section 230, Facebook's whole business model is suddenly all wrong. The company would have to break up into its different business units or go out of business . . .

Let's start by clearly delineating between "platforms" and "publishers." If a company wants to create an open forum or platform that adheres in good faith to a First Amendment standard of free speech and expression, they can do that. If they want to selectively edit and present a particular point of view and be a publisher, they are free to do that as well.

Excerpts from October 3rd 2020 medium.com article

Does Facebook censor too much or too little?

In the last year Facebook has come under attack from both sides of the political spectrum. Conservatives claim they are being censored whilst Liberals claim Facebook isn't doing enough to censor hate speech and extremism.

I worked as a content moderator at Cognizant on behalf of Facebook for nearly 2 years. What I noticed is that many types of political discourse and election materials were being monitored and censored. **Regardless of your**

ideology, we need to realize that demanding more censorship from Facebook is a slippery slope that can be weaponized by whoever holds the reins of power.

The same week that I went public as a whistleblower at Facebook on June 25th, 2020, there was a massive call for a boycott against Facebook. Many advertisers claim Facebook wasn't doing enough to bar hateful content in ads. This was part of the #StopHateForProfit boycott campaign. https://www.cnet.com/news/facebook-will-bar-more-hateful-content-in-ads-as-boycott-picks-up-steam/.

What I discovered while filming with a hidden camera at Facebook led to a criminal referral to the DOJ for Mark Zuckerberg. In April of 2018 Zuckerberg testified that Facebook doesn't censor political speech, but the evidence I presented to Congressman Matt Gaetz shows the opposite. https://gaetz.house.gov/media/press-releases/congressman-matt-gaetz-files-criminal-referral-against-facebook-ceo-mark.

I presented evidence dating back to 2017 that shows Facebook labeling many speeches from Trump as hate speech. I also showed how Facebook's policy team allowed attacks against straight white males for not supporting the LGBTQ movement. Additionally, I showed Facebook's stance on abortion and that they made "newsworthy exceptions" to promote pro-choice ideology. I have more than 20 examples of similar types of actions that favor leftist ideology and/or act against right-wing viewpoints.

I find it shocking that despite filming for 200+ hours and giving hard evidence of Facebook's misdeeds, the mainstream media continues to turn a blind eye to political censorship that is damaging a free and open Internet . . .

Section 230 of the Communications Decency Act of 1996 frames its legislation in the following manner:

> "(A)(3) The Internet and other interactive computer services offer a forum for a **true diversity of political discourse**, unique opportunities for cultural development, and myriad avenues for intellectual activity."

(2) Civil liability No provider or user of an interactive computer service shall be held liable on account of —

(A) any action voluntarily taken in good faith to restrict access to or availability of material that the provider or user considers to be obscene, lewd, lascivious, filthy, excessively violent, harassing, or otherwise objectionable, whether or not such material is constitutionally protected; or

(B) any action taken to enable or make available to information content providers or others the technical means to restrict access to material described in paragraph (1).[1]

(3) Information content provider
The term "information content provider" means any person or entity that is responsible, in whole or in part, for the creation or development of information provided through the Internet or any other interactive computer service.

https://www.law.cornell.edu/uscode/text/47/230.

The law here doesn't give a limit to how much Facebook is allowed to censor. This is what needs to be corrected and modified in this law. There needs to be a limit to how much Facebook can constrain our speech. We have the evidence that they are already censoring political speech. They are not allowing for "true diversity of political discourse."

Additionally, by deleting viral videos and promoting certain ideologies, Facebook has stopped acting as an "interactive computer service" and has now adopted the role of an "information content provider."

I last worked at Facebook 7 months ago, yet I hope the evidence I uncovered will help shine some light on the current dilemma we face between protecting free speech while limiting real-world harm.

To summarize, both sides of the political aisle attack Facebook for its misdeeds. As a free society we should only limit speech to a very small degree. Yes, there are evils in society, but differences in opinion aren't evil. **Additionally, political censorship is a tool that can be misused by both sides of the ideological spectrum.**

I hope the mainstream media will cover the facts I have presented, however it's doubtful. In a world full of people like Cat Zakrzewski, Maggie Astor, and Jayson Blair, it is hard to find truth in journalism.

Sincerely,
Ryan Hartwig

CHAPTER EIGHTEEN

The Civic Audit of Facebook by John Kyl and Covington Law

In May 2018, Facebook committed to a civic audit with former Senator Jon Kyl and the law firm Covington & Burling LLP. In August 2019, the firm concluded their year-long civic audit of Facebook. The eight-page report was inconclusive. In fact, it was barely worth the paper it was printed on.

You do a year-long investigation and come up with eight pages? What is this, an end of the semester high school essay?

Problem #1: Facebook merely listened, but there weren't enough hard data involved in the decision making.

> It doesn't include any real quantitative assessment of bias. There are no sta-
> tistics assessing the millions of moderation decisions that Facebook and
> Instagram make each day.[1]

I agree with the point here, but the *Slate* reporter's following section of the article didn't age very well at all:

> Despite the time and energy invested, the conspicuous absence of evidence
> within the audit suggests what many media researchers already knew:
> Allegations of political bias are political theater.[2]

Senator Josh Hawley said the following about this audit:

"Merely asking somebody to listen to conservatives' concerns isn't an 'audit,' it's a smokescreen disguised as a solution," Hawley charged. "Facebook should conduct an actual audit by giving a trusted third party access to its algorithm, its key documents, and its content moderation protocols. Then Facebook should release the results to the public."[3]

Were the Changes Facebook Made a Smokescreen?

After the civic audit with the Covington law firm, I noticed a few things did change. First, Facebook began tracking the exceptions. The exceptions they gave were now numbered and listed. They also stopped using the word *newsworthy* to describe their exceptions.

However, they used new terminology to allow Facebook to make decisions about how to action content. For example, the word we would often throw around would be *align*. We need all the content moderators to "align" on the same decision. This is so our scores wouldn't be negatively influenced and so that Quality Assurance (QA) and the reps were on the same page.

Yes, it's necessary to align because the policy is very nuanced. However, if there are gray areas or, as they call them, "edge cases," a decision is made from above. So Facebook could call it something else, but they could still dictate to us how to action particular jobs under the guise of alignment.

Problem #2: The Proactive Pull

When I raised up multiple posts naming the Ukraine Whistleblower by name along with his picture, I didn't know what to expect. I knew it was a big story and it was the basis by which Trump faced impeachment trials.

We decided at a local level to delete the post under the privacy policy, which I didn't agree with, because this whistleblower wasn't undercover and he wasn't law enforcement, both of which were requirements to delete under this policy.

After five hours, Facebook gave us official guidance to continue deleting the post, but under Coordinating Harm > Other, which is a policy that deals with election fraud and voter interference.

Facebook then initiated a "proactive pull" and scoured the Internet for any mention of Ciaramella, so we could delete those posts. Facebook did something similar for Greta Thunberg when she was being attacked and called "gretarded." They proactively injected classifiers into the AI, which caused those jobs to show up in our queue. This is what's referred to as a proactive pull.

On Friday, November 8, 2020, I asked Shawn Browder, the Phoenix Policy Manager for Cognizant, about the Eric Ciaramella guidance. Shawn has engaged with Facebook via teleconference and in-person meetings on a regular basis for the last two years. This is what he said:

> **Ryan:** Hey Shawn, dude like I've been getting a ton of those jobs this morning from the Ciaramella, I don't even know how to pronounce his name but—
>
> **Shawn:** It's probably because uh, Facebook's classifiers are actively pulling the content into the queues.[4]

Facebook can prioritize which type of content moderators review, through this method. The work of thousands of moderators and their reviews isn't random; Facebook controls which type of content receives priority. For a week, I reviewed hundreds of jobs related to Greta Thunberg, and on another occasion jobs related to Eric Ciaramella. I reviewed these jobs because Facebook decided protecting these people was more important than other tasks. This is another example of how Facebook can implement their bias and prioritize protecting certain people over others.

To summarize, Facebook prioritized deleting jobs attacking Greta Thunberg and Eric Ciaramella and also gave exceptions for Don Lemon (white males are terror threats) and Alyssa Milano (men shouldn't be allowed to make laws about women's bodies). The exceptions for Lemon and Milano were both for the hate speech policy, which these two individuals violated.

Solution #1: Admit Bias and Use a Sliding Scale

One solution would be to stop hiding behind the pretense of objectivity. We are humans and are essentially flawed. We grow up with our own notions, preconceptions, and biases, which are influenced by thousands of experiences, our upbringing, and our values.

As content moderators, we were always told to use our heads and not our hearts. But our hearts are what makes us human. We aren't robots, and as you can see in the Project Veritas exposé I was featured in, content moderators are clearly not objective.

One extreme experiment would be to remove the written policy completely and instead have each job actioned and decided upon using a sliding scale. On one end would be "most violating," and on the opposite end would be "least violating." Is calling for immigrants to leave a country really as bad as bestiality? Is calling someone a bitch really as bad as a video of a beheading? Yet all these examples are treated the same. They are all deleted.

By using what I call "comparative decision making," we can judge things on this sliding scale. The user could also choose which kind of content they would like to see that day, based on the same scale between "least violating" and "most violating."

This of course would only work if Facebook chose even smaller cross-sections of individuals, or even chose people based on their ideology, in order to create a broad spectrum of viewpoints.

Solution #2: Remove Facebook's protection under the Communications Decency Act, Section 230

Section 230 was legislation written in the 1990s, and since then we've seen tremendous leaps in technology and Internet infrastructure. Facebook is now using this legislation to profit off public discourse while censoring political viewpoints with which they disagree. Facebook has a poor track record that involves violating users' privacy and selling data, and now it's clear they're selectively enforcing their policy. This leads us to the inevitable conclusion that Facebook was acting as a publisher.

Therefore, if Facebook were to lose this protection, they could continue operating as a platform with conditional immunity:

> The company would have two choices: either pay a set amount of damages to the plaintiffs but continue operating as a platform with conditional immunity, or waive the immunity by declaring publisher status, dismiss the cases at hand, and then be faced with the same liability news media have to endure — risking an avalanche of defamation lawsuits naming the company as a defendant because of speech published by the company's users.[5]

> If Internet service providers are held liable for content published by their users, this could either: drive companies into bankruptcy from constant legal battles or incentivize companies to heavily regulate and censure content published by users.[6]

CHAPTER NINETEEN

Human Resources from Hell: Punishment for Exercising My First-Amendment Rights

When my coworker "Todd" reported me in early January for sharing a link to an *LA Times* article, it changed the workplace dynamic quite a bit. Around the first week of January 2020, I shared a link to an *LA Times* article about an anti-Islam protest at a mosque in Phoenix in which I participated.

For the record, the event I attended in May 2015 was marketed to me as a Freedom of Speech Rally. And to put this in context, we had had two recent terror attacks and one that originated from Phoenix. In January 2015, we had the Charlie Hebdo attacks,[1] where an entire office of satirists and cartoonists was murdered in cold blood at the hands of violent radical Islamic extremists. The cartoonists' crimes were drawing Mohammed and mocking certain aspects of Islam.

Then, in May 2015, two individuals from Phoenix traveled to Garland, Texas, in order to murder other artists such as Bosch Fawstin, who drew pictures of Mohammed. There was an art gallery event in Garland, and other anti-Islam advocates such as Pamela Geller were in attendance. These two individuals attempted the attack and were killed by police before they could achieve their goal.[2] These two men happened to go to church at a mosque in Phoenix.

It's within the context of these two events that I attended a Freedom of Speech Rally outside the mosque in Phoenix, on Friday May 29, 2015.

Attendees were encouraged to peacefully open-carry, which I did. What I later found is that "Todd," who reported me for sharing this link, also attended the same rally, but he was a counterprotester.

Oddly enough, the rally was very peaceful until the counterprotesters showed up. There was a large police presence, as well.

I was holding a sign with some of the cartoons drawn by the slain cartoonists from Charlie Hebdo. And that is the image you see of me on the cover of the *LA Times*.[3] Fast-forwarding to when I met with HR in late January 2020, I happened to be filming with a hidden camera that day. I was completely taken by surprise and had no clue that I would be meeting with HR.

Natasha Neiss is the HR rep who talked to me, and, coincidentally enough, we attended the same high school in Mesa, Arizona. I knew of her vaguely but didn't know her in high school.

She explained to me that the link I shared to the *LA Times* was "discriminatory in nature" and advocated violence:

Natasha Neiss original email to Ryan Hartwig

Date of email: Wednesday, January 22, 2020 9:50am
"To: Ryan Hartwig ryan.hartwig@cognizant.com
CC: Rodriguez, Mario* Alvarez, Jessica†

Subject: Written Warning – AUP Violation

Good Morning Ryan,

Thank you for meeting with Jessica and myself. Attached you will find the Written Warning for AUP violation.

Please review, sign and send back to me, if you want to add any comments, please add them in the email that you send back to me.

In addition, we spoke about you having 5 days to complete the AUP training on Cognizant Learn using the Code [BQVCA2_V1]

If you have any questions, please let me know.

---

* A pseudonym.
† A pseudonym.

Thank you,
Natasha Neiss
HR Talent Manager- Digital Ops
Cognizant"[4]

She also said that I violated the AUP (Acceptable Use Policy) by sharing a link of a personal nature.

Corrective Action PDF

Associate/EID Ryan Hartwig 689343
Leader/EID Mario Rodriguez‡ [employee ID redacted]
Associate Title: Process Executive
Business Unit: CDO—Social Media
Date of Issue: 1/22/2020
Issue: AUP Policy Violation
Written Warning

Brief Summary of the Issue and Expectations

Based on a thorough review, it was determined that you voluntarily sent an article that was discriminatory in nature to other co-workers within Cognizant's network. Per Cognizant's Acceptable Use Policy, user must NOT make any use or perform any activity that, in Cognizant's reasonable judgment, involves, facilitates, or attempts any of the following:

- Sending discriminatory messages or images on the Internet, via Cognizant or the client's email service or on internal Cognizant or client hosted environment.
- Advocating or encouraging violence against any government, organization, group, individual or property, or providing instruction, information, or assistance in causing or carrying out such violence, regardless of whether such activity is unlawful.

Incidental personal use of information technology assets is permitted, unless explicitly prohibited by the Cognizant client or by the specific Cognizant organization.

---

‡ A pseudonym.

Going forward, you have a duty to act to ensure compliance with Company policies, and our Code of Conduct and Core Values. It is expected that you will:

- Ensure that all use of Cognizant and client equipment and information conforms with the Acceptable Use Policy(AUP)
- Demonstrate good decision-making skills and judgement[5]

You must understand that other workers would constantly use Facebook's SRT chat system for personal use. They had their own little group chats where they would talk about random things. This was a constant trend, and leadership knew about it.

I, on the other hand, literally just shared a link. I didn't publish the *LA Times* article. I didn't say anything against Islam. I'm just holding a sign at a protest against radical Islamic extremism and the murder of cartoonists. What a radical position!

In response, I wrote a detailed letter back to Human Resources. I asked them to reverse the Corrective Action. Here is the letter I wrote back to HR:

LA Times Article Ryan Hartwig

Facts & Timeline

Overview

On January 3rd, 2020, Ryan Hartwig shared a link to an *LA Times* article from 2015. Ryan is shown in the cover photo holding a sign with various cartoons on it. Among those cartoons is on drawn by one of the French cartoonists killed in a terrorist attack (Charlie Hebdo publication).

Human Resources claims that what Ryan shared violated the Acceptable Use Policy and is inappropriate, discriminatory in nature, and advocates or encourages violence.

Please clarify what aspects of the content Ryan sent are discriminatory, who he is inciting violence against, and how attending a freedom of speech rally is inappropriate.

Ryan Hartwig was not inciting violence in any fashion. He was holding a sign that contained no violent imagery.

Link to article
https://www.latimes.com/nation/la-na-anti-islam-demonstration-20150529-story.html

Timeline
-1/3/2020 Ryan Hartwig shares link to Daniel Will and Rex Thomas[§]

-1/22/2020 Ryan Hartwig meeting with HR (Natasha Neiss) and Jessica Alvarez[¶] (TM) –Corrective Action issue.

-1/22/2020 Ryan Hartwig signs CA, completes AUP re-training.

-1/24/2020 Ryan Hartwig emails Natasha, asking her to re-visit the decision and gives new information about how Rex Thomas allegedly re-shared the link using the same client messaging device. Rex Thomas allegedly libeled Ryan Hartwig or slandered Ryan Hartwig while communicating with other co-workers about Ryan Hartwig.

-1/28/2020 Ryan Hartwig includes Jim Denton[**] (site leader) and Lexi McPherson[††] (deputy manager) in the conversation, has not heard back from Natasha, and asks for further clarification. Ryan Hartwig requests that the corrective action be removed.

-1/28/2020 Ryan Hartwig emails the Chief People officer James Rhenquist[‡‡], who has responded personally to Ryan's emails in the past. Ryan also emails the Global PR email since this deals with 1st amendment issues.

-1/28/2020 Lexi McPherson replies and says that Jim Denton and her will look into it.

-1/28/2020 Natasha Neiss replies back and reaffirms the decision for the Corrective Action, and adds that the content I sent was inappropriate and violated the AUP, and that the decision will not be reversed. She offers to meet up to explain further and I accept that offer and reply that I want to meet with her.

Facts

- The link I shared was an article that mischaracterized my role as an individual engaging in and exercising my 1st amendment rights to freedom of speech.
- Rex Thomas admitted to me on 1/3/2020 that he was in attendance at the same Freedom of Speech rally as a counter-protester, and conveyed a statement of hostility towards the protesters.
- Supervisor (SME) Jason Ngo§§ told me that he was shown the *LA Times* article, and it wasn't shown to him by me, nor did I send him a chat with the article.
- There are continuous violations of the AUP occurring on a continual basis amongst employees on this project. Also, in Mario's Team chat there are reoccurring and continual examples of insensitive jokes directed towards groups of people.
- Employees on this project are continuously streaming video game content and watching movies while at work. They are able to create their own group chats with innocuous names such as "wellness chat" where they often share insensitive content.
- This project ends on February 29, 2020.¶¶

Other Issues/Questions

- Is there an arbitration agreement?
- I've raised up issues to Natasha Neiss before, but never have I been the subject of a review for discipline. Since Natasha Neiss and I attended the same high school in Mesa 2 years apart [name of high school redacted], this may be seen as possible conflict of interest. I didn't directly associate with Natasha Neiss during high school, but I knew of her.

§§ A pseudonym.
¶¶ This is referring to the fact that that Cognizant/Facebook business relationship would end on this date and we were all about to get laid off.

Questions for Cognizant

Does Cognizant punish employees who share examples of their exercise of their 1st Amendment rights?

What is Cognizant's stance on the 2015 Charlie Hebdo attacks, where multiple French cartoonists were murdered for portraying the Islam religion in a satirical nature? Cognizant Employee Ryan Hartwig was holding a cartoon drawn by one of the victims and was protesting their senseless murder.[6]

As you can see, I asked them some very tough questions in this letter and made some solid points, but they still refused to reverse the corrective action.

However, this letter did lead to a second meeting with Jim Denton (site lead) and another HR supervisor. Here is one of my other email responses:

In an attempt to defame my character, the original link I shared was distributed amongst other individuals, including a supervisor.

The individual who may have reported this was a participant in a counter protest at the same event in 2015.

I understand there can't be retaliation against the original reporter, but if this link was shared using the same device in order to libel me or defame my character, then I would request that the Corrective Action be removed.

The link I shared was an article that mischaracterized my role as an individual engaging in and exercising my 1st amendment rights to freedom of speech.

Please escalate this to the legal corporate team to avoid a possible lawsuit against the individuals who defamed me for exercising my 1st amendment rights. Please reverse the Corrective Action.

Thank you,
Ryan Hartwig
689343"[7]

Consequences of the Corrective Action

Having a CA (Corrective Action) wouldn't immediately disqualify me from a transfer to another project within the company, but it would be taken into account. After working there for two years and only missing work two or three times, I felt Cognizant was treating me unfairly.

In addition to sending my letter to Human Resources and local leadership, I also emailed the global Public Relations lead. It's ironic because if they had reversed the corrective action, I would have had less of a story to present to the public. And now we see how this whole issue has the potential to become a very large PR fire for Cognizant.

The local HR rep Natasha said that because of how the complaint was initially sent, her hands were tied as far as reversing the corrective action. The HR complaint was sent directly to HR compliance, which is like HR corporate.

This is the reason she gave for not being able to reverse the corrective action.

Another fascinating aspect of this scenario is that Cognizant could not access my chat messages directly. They had to ask Facebook for permission for those data. That means someone at Facebook may have reviewed my messages and then sent them to Cognizant.

It's hard to know if Facebook played a role in the decision to punish me for attending a protest.

But it's not outside the realm of possibility.

Facebook was the client, and ultimately, we answered to them. I was using the SRT (single review tool) Workplace Facebook chat, which is a proprietary tool belonging to Facebook.

This whole incident with HR was documented in the original video exposé with Project Veritas, which was released on June 25, 2020. In the same exposé, we see my coworker Kassi Cimo openly talking about taking the eighty-million-dollar bounty from Iran to assassinate Trump. It's incredibly ironic that she did NOT get in trouble with HR for talking about killing the president, yet I got in trouble for holding a sign at a protest five years prior.

I have many friends who are Muslim. When I attended Arizona State University, I made friends with many Muslims. ASU is the eighteenth most diverse university in the United States, and I even attended the mosque near campus for Friday prayer on a few occasions. I also took Arabic 101 during my last year of college in 2014–2015, and I know a few phrases. So by no means am I a bigot or a racist. What I am concerned about are the radical elements within the religion.

To summarize, I attended a free speech rally in 2015 that was protesting the murder of cartoonists by radical Islamic extremists. If there were radical Christian extremists shooting up productions of *The Book of Mormon*, I would protest that, as well. However, the depth and breadth of my involvement with Islam is much more than one protest. I've even worshipped with Muslims on Friday prayer in Tempe, Arizona, and learned about their culture from newly arrived exchange students.

Every Muslim I have known is a kind, generous, and humble soul. I have nothing but the greatest respect for the Islamic faith, but when you shoot and kill a bunch of cartoonists, I'm going to stand up.

In my opinion, Human Resources mishandled the situation from the beginning, and they punished me for sharing a link to a news article. They completely mischaracterized my behavior and never reversed the corrective action, although they did modify the rationale for the reprimand.

I do believe Natasha Neiss was put in a tough situation, mainly because the initial complaint was submitted above her chain of command. However, once Ralph and the other Human Resources personnel were involved, I feel that they could have done more to reverse the Corrective Action[***]. Once again, the publicity blowback for them was more severe because of their refusal to change the punishment.

[***] The Corrective Action (CA) was an initial written warning, so it wasn't a fireable offense, but it did limit my possibilities of transferring to a new project within the company and would stay on my employee record.

CHAPTER TWENTY

Preparing for the Main Event

In May 2019, I'd composed my letter with multiple instances of alleged bias by Cognizant against conservatives and sent it out to several congressmen and United States senators.

None of them responded.

After a few weeks, I sent the same letter to several conservative news outlets. Someone from Breitbart News responded and suggested I should contact Project Veritas, which I did.

Project Veritas responded immediately. I talked to an employee about what I'd observed, and they helped me get a camera to record some undercover video. I don't want to talk about the camera and the technology employed, but those first few times I filmed in June 2019 I was terrified that I'd be discovered. Quite honestly, for the next few months I didn't film much.

I talked to Project Veritas in October 2019 and decided to resume my filming. In November 2019, I received word that Cognizant would be ending its contract with Facebook at the end of February 2020. That gave my work an added sense of urgency. Cognizant might be exiting the content moderation business, but "the client" would still be very much in the game.

When it was announced that Cognizant was leaving content moderation in late 2019, there was a lot of discussion about why they made the decision. The company line was they wanted to concentrate on their core business, which was cloud computing. As employees, we couldn't help but believe that the negative publicity* about the toll content moderation that was taking on some of the employees was also a significant factor in their decision.

* * *

When I first started filming in June 2019, I thought it would be interesting to focus on how Facebook was framing racial issues, specifically the "white people are bad" narrative.

In March 2019 there was a shooting at a mosque in New Zealand, and that was a perfect fit to Facebook's narrative about the danger of white males. While my initial angle was to focus on how Facebook framed racial issues, I decided the better angle would be to focus on what Cognizant was doing at Facebook's direction to actively censor conservative voices.

It wouldn't have been enough to show that content moderators were overwhelmingly liberal. The public had a right to know that the actions they took were designed to silence those on the right from the public conversation. The public deserved evidence about what Facebook was doing with their content moderation policies.

I was somewhat frustrated by the fact that I didn't deal directly with Facebook. However, Shawn Browder, our policy manager, did interact with them. Shawn actively made himself available to us, often taking a seat at a desk just at the edge of the production floor so we could ask him questions. Even though I didn't interact directly with Facebook, I made sure to direct many questions to Shawn, who was the main point of contact with Facebook for us to implement their policies. I'd prepare my questions for him, then get up from my desk and walk over to him.

I might say, "Hey Shawn, I just need a little more clarification on this Greta Thunberg thing. Why is she getting more protection than other public figures?" He'd give me the rationale, then might add a little bit more detail and say something like, "Yeah, they're using the A.I. [Artificial Intelligence] to scrape the Internet for negative mentions of her and remove them." I'd nod, taking care to conceal my shock at this revelation, then as I walked to my desk think to myself, *Good God, they're using artificial intelligence to remove any criticisms of Greta Thunberg from the entire Internet? How is that possible? Is that even legal? Hey, can I use that service, so I make sure nobody ever says anything bad about me on social media?*

* The negative publicity was generated as a result of the February 2019 article published by "The Verge" titled "The Trauma Floor: The secret lives of Facebook moderators in America." Before this article was published, the general public didn't even know that Cognizant was doing this type of work.

When it was announced Cognizant was leaving the content moderation business, there was first the inevitable shock that we'd soon be looking for another job. But after that passed, I felt the atmosphere lighten. The pressure was off, people seemed to slow down, and the employees were more willing to talk to one another. It was a great filming opportunity, and I recorded a lot of political conversations.

My wife, Livy, found out I was filming in early June 2019, and we discussed it. She considers herself politically neutral, and we talk often about politics. But she was worried about my safety, and nervous about losing Cognizant's good health benefits. I told her I'd be careful, and to her immense credit, she understood the importance of what I was doing.

* * *

By January 2020, I'd probably recorded more than a hundred hours of undercover video. But that wasn't enough, as I needed something more, especially with Cognizant getting ready to exit the content moderation business.

The evidence was overwhelming. I needed to nail "the client."

Because I look the way I do, and it was clear I didn't share many of the liberal beliefs of the typical content moderators, I needed a way to extract more information. I thought that your typical liberal won't be honest about what they feel if they suspect you of being a conservative, so I wanted to change that dynamic.

Project Veritas sent out a few undercover journalists who built rapport with my coworkers and talked to them about content moderation.

The two employees in the video release who talked to the undercover journalists were Jose Moreno and Tyrell Lease, two military veterans who were conservative and didn't care if anybody knew it. At work I'd made a special effort to sit near them. I was concerned about how they'd react to being on the Project Veritas video. But when the video dropped, they were thrilled because they wanted Facebook outed just as much as I did. Jose texted me after the video came out and said something like, "Dude, you guys did a really good job."

One of the undercover journalists talked to another coworker, whose face was blurred, and she admitted to always deleting MAGA and pro-Trump content.

There were some troubling interactions, as well. One of my coworkers gave us some extraordinary content but after several drinks talked about how he was often suicidal. I knew that coworker had often been seeing the counselors at work for both work and personal problems. I told the journalists we couldn't show him in the video because of his fragile psychological state. They agreed and kept him out of the video.

It was a late night for the undercover journalists, who were able to gain useful information about some of the practices and attitudes of my coworkers and which actions they took to censor conservative content.

* * *

The Project Veritas journalists stayed for several more days and interacted with my coworkers on several occasions outside of the workplace The final Friday in February several employees hung out at a bar across the street from Cognizant to commiserate about the ending of the job. There was some talk about bias, but it was mostly about what they were going to do next.

Cognizant gave us a severance package of about five thousand dollars, which I took. In the subsequent interviews I've done, the host always seems to ask me about being "fired" by Cognizant, and I have to point out to them that that never happened.

The job ended because Cognizant was no longer working with Facebook. I never got "caught."

In early March 2020, a couple of weeks after my job at Cognizant ended, Project Veritas flew Livy and me out to New York to film my interview with James O'Keefe. We stayed at a hotel near New York City and my uncle lives near there, so we spent some time walking around the city with him.

My interview with James O'Keefe took place in a studio in New York City, rather than their studio in Mamaroneck. James is a tall man, broad-shouldered, with a cherubic, boy-next-door look and charm to him.

It was in the studio before my interview that James first told me about the other Facebook/Cognizant whistleblower, Zach McElroy, who worked at the Cognizant facility in Tampa, Florida. I was shocked. Project Veritas really liked to play it close to the vest. But I quickly understood the need for tight security when you're going after one of the biggest companies on the planet.

The interview flew by quickly, and I did the interview with James in a single take. James and I were both dressed in a suit and tie, and I remember this curious feeling of calm descending on me as the interview began. This was a righteous action I was taking to protect free speech on social media, and my soul felt joyous about what I was doing.

The COVID-19 crisis was beginning to dominate the news with talk of lockdowns, but there were still no restrictions in that part of New York by the time we left. Ironically, it later turned out we were near one of the virus hotspots, but Livy and I never came down with an infection.

While the five-thousand-dollar severance package from Cognizant was a great help, that would last maybe a month and a half, and I'd need another job. My job with Cognizant ended at the end of February 2020, and within four weeks I was working for a home warranty repair company fielding calls from people who needed their appliances fixed. Summer tends to be the busy season for appliances to break, so they were staffing up for the expected surge. Although the job was forty hours a week and I could work from home (because of the COVID-19 lockdown), there was also likely to be significant overtime available to us.

The plan had been for the Facebook/Cognizant whistleblower video with me and Zach McElroy to premiere in late March or early April. However, it kept being put off because of the COVID-19 crisis in the news cycle. I was given three different dates in June when the video was supposed to drop, until it finally did on June 25.

June 25 was a Thursday, so I took that day and Friday off work, knowing I'd be doing a number of interviews. I also anticipated I'd be leaving my job, so I set up a GoFundMe page, and within that first week I raised a little more than fifteen thousand dollars, which would allow me to dedicate a few months to pursue the issue. The day of the release I did a number of interviews, feeling like a deer in the headlights, but by the second day I felt as if I were finding my footing.

The Monday after the release, I went to my boss at the home repair warranty company and told him I was quitting because I needed to be free to pursue this issue for a few months, before figuring out what I wanted to do next.

However, as prepared as I thought I was, I didn't realize how much the Project Veritas video would change the course of my life. There would be no return to anonymity for me.

CHAPTER TWENTY-ONE

Project Veritas Goes Live

On June 25, 2020, my months of recording and transcribing undercover conversations and documents were all coming to a head.

Project Veritas likes to put together fifteen-to-twenty-minute "video packages," almost like *60 Minutes*, abundantly supported by documents and video. I wasn't the first Facebook whistleblower for Project Veritas. Shortly before my video aired, Zach McElroy, another content moderator for Facebook who worked for Cognizant in Florida, was featured by Project Veritas.

Production quality is high, and as I understand video, most of the time you're changing images every few seconds, so the eye doesn't get bored. On the morning of June 25, 2020, I got up early, checked my email, and saw the video had come out. With great trepidation, I clicked the link and began to watch.

The first thing I saw were some quick graphics of Facebook and technology, then a voice said, "Facebook favors the left. They do. 100 percent."

Another voice said, "We work with a lot of liberals. Yeah."

A third voice claimed, "If I wore a MAGA shirt or MAGA hat, I'd get my ass beat."

Then on the screen a title appeared, "FACEBOOK INSIDER SPEAKS OUT," video of the six-story Cognizant building in Phoenix, Arizona, and my voice saying, "I saw some discrepancies and I saw some evidence of bias."[1]

From the footage of the Cognizant building, the video jumped to Leslie Brown, identified as a senior human resources Partner at Facebook. An undercover journalist is asking her about the ease of firing a white guy. "You're saying, because he's a white male, oh, it's easier when they're white men?"

"No one has the white man's back anymore," she replied and laughed.

The video jumped to another recording where the guy says, "We rig the game so that it could work on the left side."

My image appears for the first time, short blond hair, blue eyes, glasses, wearing a suit, and yes, a white man. I said, "I saw more blatant posts against Trump."

The video jumped to another undercover journalist who asked the subject, "But Facebook obviously f**cking hates Trump?"

The subject, this time identified as Jose Moreno, content moderator – Facebook (Cognizant), replies, "Yes. Yes. 100 percent, they do."[2]

The next image was James and I sitting in the Project Veritas studio, having a conversation. James is tall, dressed in a suit and tie like me, handsome with a youthful, cherubic face, but deadly serious when it comes to corruption and hypocrisy. He's nodding as I explain, "I saw an alarming number of posts that really focused on conservatives. Kind of a double standard."

The video cut to an undercover video of an individual identified as Steve Grimmett, team lead for Content Review – Facebook (Accenture). He was outside the Accenture building in Florida, the camera angle odd from what seems to be a hidden camera at waist height, but it framed him against the blue sky and fluffy white clouds. He says, "So, I've spent quite a bit of time looking at pictures of hate organizations. Hitler, Nazis, MAGA, you know, Proud Boys, all day long."[3]

While watching, I'm imagining how viewers are taking in these claims. Yes, I'd heard similar things in the nearly two years I worked at Cognizant, but I'm visualizing how it's striking the millions of people I know will be watching this video in the coming days. The video cuts to James with the first Facebook insider, and his voice said, "Our latest Facebook insider, Zachary McElroy, exposed the pervasive anticonservative bias in Facebook."

James had recorded video of him and me walking into the studio. I thought it was corny at the time, but it worked as his narration of the scene began, "Today, another Facebook insider, Ryan Hartwig, from Phoenix, Arizona, comes forward. And he says the anticonservative bias is pervasive." The camera focused on James and me sitting down as we began a longer conversation. The sizzle of getting people interested in the segment was done, and now we were getting to the steak.

James: Tell us your name and where you work.

Ryan: My name is Ryan Hartwig, and I work as a, or as of a couple weeks ago, as a subcontractor for Facebook, for Cognizant, in Phoenix, Arizona. I was a content moderator for Facebook, essentially.

James: And why did you come to Project Veritas?

Ryan: I thought people deserved to know what was going on. Seeing such blatant bias from Facebook just really bothered me.

James: Other than your personal experiences, you felt there was an institutional bias that you say was aggressive at Facebook. Can you talk about that for a minute?

Ryan: Yeah, so when I started in March of 2018, I saw more blatant posts against Trump. I saw an alarming number of posts that really focused on conservatives. Kind of a double standard. There were like six people who decided policy for all of Facebook and they all think the same. They're all very likeminded. So, if you live in San Francisco, you're not going to find a sample population that's diverse in their political viewpoints.[4]

The footage jumped to a video recorded on February 20, 2020, with Jose Moreno, a content moderator at Cognizant. Jose was wearing a green-and-black-striped cap, glasses, and had a short beard and mustache. With the cap coming to something of a point on his head, I thought he looked a little like one of Santa's elves.

Jose: We rig the game so that it could work on the left side.

Journalist: How?

Jose: Uh, freedom of speech is the main one.

Journalist: So, they're allowing political ads still?"

Jose: Yes.

Journalist: So, now they can get more exposure to the left versus the right?

Jose: Yes.

Journalist: But Facebook obviously hates f**cking Trump?

Jose: Yes! Yes! 100 percent they do.[5]

It went to some of the footage I'd shot inside the Cognizant office. I was speaking with Israel Amparan, a fellow content moderator. Israel has dark hair flecked with gray, glasses, a goatee, and an intense manner about him. Hearing my voice for the first time in the segment on undercover video was both exciting and nerve-racking.

Ryan: But, yeah, we've been getting a lot of content about it. I've had at least ten jobs today.

Israel: A bunch of s**tty f**cking rednecks threatening civil war if they remove 'our duly elected president.' But Trump supporters are f**cking crazy ass assholes. Every other f**cking word that comes out of their mouth is, you know, "come take it," "civil war," and things like that. And like listen, I hate government as much as the next f**cking person. But you're not going to catch me rioting over the f**cking—it's like impeachment. It's like, that's a problem. It doesn't just happen. Trump called it a f**cking coup. And it's like that should scare you more than anything.[6]

From Israel the focus was back on me as James continued his narration, saying, "Hartwig was not surprised when content moderators at Facebook that he worked with grouped average Joe, Trump supporters with terrorists." The camera angle switched to a wide shot of me and James sitting in the studio. Between us James had a large video monitor set up so that he could play clips and have me respond to them:

James: The next clip was captured by hidden camera. This was Steve Grimmett, team lead for Content Review, Facebook (Accenture). [Grimmett was talking to an undercover Project Veritas journalist. James started the video clip.]

Journalist: I guess that maybe part of your job description is looking for red flags. But you know, sometimes they're redder than others.

Steve: One of my projects before now was hate. So, I've spent quite a bit of time looking at pictures of hate organizations, Hitler, Nazis, MAGA, you know, Proud Boys. All that stuff. All day long.

James ended the clip and then looked at me.

James: Does it surprise you that he combined Hitler, Nazis, and MAGA? He's describing hate organizations he's moderated for Facebook, and he just throws MAGA in there. What is your reaction to that?

Ryan: He groups Hitler, Nazis, MAGA together and that's how the moderators are conditioned to think. Like, hey, anything that's right wing, it could possibly be on the hate list. There's no left-leaning individuals on that list. The policy is called "Dangerous Individuals and Organizations." On that same list are terrorists.

James: One of the moderators you mentioned, Shawn Browder, was a [Bernie] Sanders supporter. [Actually, Shawn Browder was in charge of moderators.] How about the people who were making the choices as to whether to take something down? What were their politics like?

Ryan: I remember one individual named Kassi, who sat behind me to one side. We had a few conversations about politics. On one occasion, this was a couple weeks ago, a bounty was placed on Trump's head from Iran. They offered eighty million dollars to anyone who would kill President Trump. And Kassi was talking about how she'd accept that bounty. And that it would be worth it.

James: She's a content moderator?

Ryan: Yes.[7]

The video switched to my fellow content moderator, Kassi Cimo. She had shoulder-length brown hair, was of about average appearance, and in most situations was a nice person. She was troubled by the idea of handing Trump over to Iran but seemed to have made her peace.

Kassi: It's inhumane. It's inhumane. But if it's gonna save the country, why not do it? I feel like eighty million won't do a lot for the debt. I'm just saying.

We should just hand him over to them. Take the money, as a country. That's what I'm saying. If we hand him over our country would be saved. Just saying.

Ryan: Yeah. I mean, it's a bargain, right?

Kassi: I'm just saying, take him [Trump]. Y'all [Iran] can keep your eighty million or you can give it to us and we can put it into our debt. Like, you're saving the US. Like, come on. Yeah, that's it. They just want one person. Why not take one for the team?

The video switched back to images I'd recorded of the Cognizant office as James continued his narration.

James: Hartwig said there was also an alarming anti-white bias that was being applied to Facebook's content moderating policies.

The camera was now on a closer image of James and me in the interview.

James: Okay, we were talking about this LGBTQ slide. Talk to me about what we're looking at here.

Ryan: Every summer there's Pride month. This first came to my attention in the summer of 2018. I'd been there for about four months. Shawn Browder came by and gave us a policy update.

James: Who is Shawn Browder?

Ryan: Shawn Browder is the policy and training manager for Cognizant. [A photo of Shawn Browder and his title appeared next to my image.] He essentially oversees, makes decisions for policy for over a thousand Cognizant employees. So he has some autonomy. He came in and gave a policy rollout about, "Hey, this is going to be Pride month," and he was speaking to a group of mainly Hispanics, because I was on the Spanish team for the first year and a half. But this basically says, straight out, "Hey, we're making exceptions for our policy to favor the LGBTQ community." Basically, it's okay to call straight white males "filth," in the context of not supporting the LGBTQ community.

That's the specific context. It's a carve-out, a specific exception for the policy and it allows attacks on one single group of people, straight, white males for not supporting LGBTQ rights.[8]

The camera focused on the document I'd filmed on the computer screen, "Content to Raise to the Policy Team – Raising Awareness for Pride/LGBTQ Community."

James: Hate speech is allowed if it is "intended to raise awareness for Pride/LGBTQ." So, hate speech is allowed in some cases, but apparently not in others?

Ryan: Correct. Yeah.

James: We've heard this kind of anti-white bias in Big Tech before. Last year one of our undercover journalists spoke with Leslie Brown, a former HR contractor for Google, who now works as an HR executive for Facebook in San Francisco. She laughed at the anti-white bias that exists inside Big Tech.

The screen switched to Leslie Brown, sitting in a restaurant being filmed by a hidden camera.

Journalist: They were able to fire him without having to worry about discrimination?

Leslie: Due diligence, right. Because he's a white man. Yeah, white man, no problem. You can't do it that easily if there are other issues.

Journalist: Oh, it's easier then they're—

Leslie: White men. No one has the white man's back anymore. [Laughs.]

Journalist: You're saying because he's a white male, there was more—

Leslie: That if he chose to sue the company, that most attorneys would just laugh.[9]

The video switched back to James and me in the studio, with James reading his questions off a legal pad.

James: When you started, you felt like something was off. Do you feel you were being targeted for your politics?

Ryan: Yeah.

James: Hartwig said Facebook's anti-conservative, anti-white bias made him a target. Did you know that other employees called you racist from day one?

Ryan: No. No. And when I transferred to the North American side, I didn't hear anything about that. I didn't know people were talking about me behind my back. I never really caught wind of it. I understand, people probably knew I was conservative, or Republican, that's very possible.

James: You weren't completely alone. There were other conservatives who worked with you who also felt out of place. And tell us about some of those people.

Ryan: A few of the people I sit with, Jose Moreno and Tyrell Lease, they're both very conservative and they're both military veterans. But they noticed the content they were moderating targeted conservatives. There was bias and kind of a group-think effect that was happening, where you had a lot of liberal people, or people who were "actioning" a certain way, and you also had people who were just following the rules.[10]

The image flipped to Jose Moreno, the conservative and military veteran I sat next to at work.

Journalist: That's what you said, Facebook favors the left.

Jose: They do. 100 percent.

Journalist: Let me ask you this. Do you feel Ryan got targeted because he's conservative?

Jose: He did. 100 percent.

Journalist: Is everybody you work with mostly like you guys, or everybody's left?

Tyrell: No, we work with a lot of f**cking liberals. Three-quarters.

Journalist: So, most of the people moderating—

Tyrell: If I were to go in there with a MAGA shirt or MAGA hat, I'd get my ass beat.[11]

The narration returned to James O'Keefe.

James: Hartwig says the anti-conservative bias in his case went beyond name-calling. He was called in for "corrective action," by his HR department for what he says was a benign action. Then you post an *LA Times* article to an internal message board. Tell us about that.

Ryan: This was around early January [2020], we were having conversations about Islam and different religions and I shared a link to this article. This was back in 2015. In January of that year there was an attack in France, and the Charlie Hebdo cartoonists were murdered. There was an attempted attack in April [in Texas], and the two individuals were from, they attended a mosque in Phoenix. I learned about the event, it was labeled as a "Freedom of Speech" rally.

James: What did the Facebook HR people tell you?

Ryan: They told me, "Hey, you violated the acceptable use policy. You violated our rules. You used it for personal use." But in the document they gave me, which they never retracted, they said that I was advocating violence.[12]

In what I can only claim as divine intervention, I was wearing my hidden camera when they called me into Human Resources and filmed the entire crazy encounter. I was reprimanded by Natasha Neiss, Cognizant HR, and her image came on the screen.

Natasha: Yeah, the AUP policy states that you cannot send or use Cognizant or client internal systems to send discriminatory, racial—

Ryan: Was it the *LA Times* one? The *Los Angeles Times* where it had an image of me during a protest in 2015?

Natasha: Yeah.

Ryan: Oh, wow, okay.

Natasha: Yeah, so, especially those types of articles, whether or not you're mentioned in them, or whatever the case is. It can be offensive to other people. So, it's a violation of our AUP. You cannot use our systems to promote things like that.

Ryan: Got you.

Natasha: You still violated the AUP policy—

Ryan: Because I shared a link?

Natasha: By sharing an article that was discriminatory in nature.[13]

The video returned to James O'Keefe.

James: So you said you went back the second time and you said they backpedaled on something? What was that?

Ryan: In the initial corrective action, CA, they said I violated the AUP and I was advocating violence and the thing I shared was discriminatory in nature. And I met with them, they said it wasn't the fact it was discriminatory, no, no, no. It wasn't any of that. It was simply that you used the client device for personal use. So they reiterated it, it wasn't anything you shared, it was the fact that you shared it for personal use.

James: I see. They backpedaled. They changed their tune.

The image switched back to undercover video footage of Natasha in the second Human Resources meeting, with another employee present. I must have really scared them with my claim that my First-Amendment rights were being infringed upon, because they spoke specifically about those rights.

Natasha: So, again, you have every right to exercise your First Amendment, we are not taking that away from you. We just ask that you don't do it inside the building on our systems.

Ryan: So, I kind of understand what you're talking about. So, I guess, like I just typed it because I just wanted to have all the facts in one place. But like, so the fact that it was an *LA Times* article, or the fact it was involving anti-Islam protest, or a Freedom of Speech rally, that you're saying that's kind of beside—

Natasha: It doesn't even matter. If you sent, I'm trying to think of something else, an invite for a children's event that's happening downtown, it would still be reviewed for violation of AUP policy because you sent something non-work related using our systems. So, that's again or bottom-line issue, is that you used our systems to do non-work related stuff.[14]

The video shifted back to James, asking me a question.

James: You felt you were targeted because you were conservative, and also felt you were being passed over for jobs and discriminated against for being a white male. Tell us about that.

Ryan: I applied for the policy team twice. There's no question about my fluency in Spanish and that was part of the job requirement. But as far as being a straight white male in the workplace, yeah, the policy we saw about how straight, white males are "filth" kind of breeds that culture. Hey, we allow attacks on straight white males, and Ryan is a straight, white male. I feel like I was, perhaps, limited in my advancement.

James: I think we have some tape of people saying you were targeted because you are a white male.[15]

The next image was hidden camera footage of Gryselda Iniguez, another content moderator at Cognizant.

Gryselda: He applied for two jobs that he should have got.

Journalist: But he didn't get them?

Gryselda: Because he was white. I know because he was white.

Journalist: Ryan applied for two jobs; did not get them because he's white and conservative? Conservative or just [white]?

Gryselda: No, because he's white.

James: Ryan wasn't the only person targeted by the conservative bias at Facebook.[16]

The video shifted to a content moderator whose face was blurred, but we can clearly hear her words.

Content Moderator: I stopped being friends with Ryan and Raymond, cause we were all on the same team.

Journalist: Yeah.

Content Moderator: But they bonded over their support for Trump and they would always like bring it up and they would get mad because I would go at it with them.

The image went to another snippet from Jose Moreno, the bearded guy in a cap who was a former military veteran, and a friend of mine.

Journalist: So, Facebook's like favoring the left, with its policy?

Jose: Yes. It is.

Journalist: Okay.

Jose: And with people getting banned and blocked.[17]

We returned to James and me in the studio, sitting across from each other.

James: Most people, just don't have the, speaking Spanish, the *cojones*, to do this. I mean, I'm assuming you had a confidentiality agreement with Facebook's Cognizant and you, basically strapped a camera to yourself and recorded that training. And that training, it does say "confidential election, 2020." Are you worried?

Ryan: It was very stressful. The first couple months of recording, my wife didn't know about it. So, when it came up it became a stress point for our marriage. I signed the confidentiality agreement with Cognizant. But after that point I was seeing them interfere on a global level in elections. Then I saw a blatant exception that targeted conservatives or favored liberals. And you know, we're deleting or actioning three hundred posts a day. (Per content moderator.) So, if you magnify that by however many content moderators there are on a global scale, that's a lot of stuff that's getting taken down. So, that was the tipping point. There was that one last, that one final post that put me over the edge. But then knowing what I knew about how they were giving exceptions for the policy. I knew this was likely happening on a global scale.

The visuals changed to me walking outside the Project Veritas studio in a local neighborhood with the other Facebook whistleblower, Zach McElroy. James ends the narration:

> **James:** These brave insiders who felt compelled to come to us with this story, worked for a Facebook contractor. At the end of February, Facebook ended the contract with their employer. But the bias they exposed has not been addressed by Facebook.[18]

The camera went dark on our images and the Project Veritas logo came up, encouraging people to "Be brave. Do something."

* * *

I sat back in my chair after watching the video and just let it all wash over me. To this day I still feel a jumble of conflicting emotions when I view that video.

I think of all the people I've known throughout my life, our countless interactions where I hope that I have shown myself to be a good person, then consider millions of people being introduced to me for the first time in a video that lasts a little less than eighteen minutes. Yes, I believe I was targeted for being a white male and a conservative. But I have no desire for vengeance in my heart. Did Israel Amparan curse more in ten seconds than this Mormon has cursed in his entire life? Yes. But do I want to force him to endure a "sensitivity training" class? No, I do not.

Does Kassi Cimo understand how completely crazy it sounds to hand over any American president to Iran, a regime that throws people off buildings for the crime of being gay? Would she have been similarly understanding if somebody suggested handing Barack Obama over to some foreign power?

I am still stunned at how my sharing a link to an *LA Times* article, supporting freedom of speech in the wake of the Charlie Hebdo massacre in Paris was initially categorized as "hate speech." For those of you who may have forgotten, the Charlie Hebdo massacre was when two French-Algerian brothers killed seventeen people, because their satirical magazine had published cartoons about the prophet Mohammed. Do you understand how terrible this was? They didn't attack soldiers, or policemen, or a bank.

They went after cartoonists.

America fell in love with the satirical 2011 Broadway show *The Book of Mormon*, by Trey Parker, Robert Lopez, and Matt Stone, from the TV show *South Park*. In fact, there are still performances of the show around this country right now. Do the creators of *The Book of Mormon* worry that they might be gunned down by fanatical Mormons?

No, they do not.

I support the right to free speech, even when I disagree with it.

When did that become such an outdated idea?

Because if you say that free speech may be limited, and that we, as a society, should support such restrictions, actions like the Charlie Hebdo massacre become, if not permissible, then understandable. I want to put you back into the actual events of the Charlie Hebdo massacre. Because if we look away from it, I fear we are entering a new dark age of religious fundamentalism. This is how the attack is described in the Encyclopedia Britannica:

> On January 7 [2015] the offices of Charlie Hebdo were the target of a terror-
> ist attack. At 11:30 am Algerian French brothers Cehrif and Said Kouachi,
> armed with assault rifles, entered the magazine's offices and killed caretaker
> Frederic Boisseau. They then forced cartoonist Corinne ("Coco") Rey to enter
> the security code that granted access to the second floor, where an editorial
> meeting was being held.

> The attackers stormed into the newsroom, and police officer Franck
> Brinsolaro, who had been detailed to protect Charlie Hebdo editor Stephane
> ("Charb") Charbonnier, was shot before he had a chance to draw his weapon.
> The attackers then asked for Charbonnier and four other cartoonists, Jean
> ("Cabu") Cabut, Georges ("Wolin") Wolinski, Bernard ("Tignous") Verlhac,
> and Phillipe ("Honoré") Honoré, by name before killing them as well. The
> other victims were economist Bernard Maris and psychoanalyst Elsa Cayat,

both columnists for Charlie Hebdo, copy editor Mutsapha Ourrand, and journalist Michael Renaud, a guest at the meeting.[19]

The protest at the Phoenix mosque that I attended was preceded by a similar attempt in Texas, when a cartoon drawing contest in support of Charlie Hebdo was targeted by two individuals who attended the Phoenix mosque.

I hope I've made it clear I'm against fundamentalism, whether it comes from Islam, the left, or even the right. I don't want to force the left-wing radicals at Facebook, or whatever company is used for content moderation, to attend "sensitivity trainings" on the feelings of conservatives. But maybe a class on basic American civics wouldn't be a bad place to start. When did the left get the idea that if you have conservative ideas, that means you should not be allowed to speak? I presume goodwill from those on the left. Why don't they do the same? The answer to this problem doesn't lie in regulations, training, or policies, but from the human heart, where we grant every person the innate dignity given to them by God.

My intention in going public was never to shame or embarrass Facebook, Cognizant, or many of my former coworkers. It was to hold a mirror up to them, so they might see themselves more clearly, and allow their own sense of morality and common sense to guide them in a better direction.

CHAPTER TWENTY-TWO

The Media Merry-Go-Round

Nothing can prepare you for life as a public celebrity.

One day you're just an average Joe, looking for work because your previous employer lost their contract with Facebook. The next day your video has a couple million views, and even though you're walking around your town with a face mask on because of the COVID crisis, people start to recognize you.

I think it's probably important I discuss how the actions of a single writer, Casey Newton, destroyed Cognizant's contract with Facebook and even raised the question of whether human beings should be reviewing content at all on social media.

My Project Veritas video also had an effect, highlighting questions of media bias in content review. But to some extent the impact was blunted because Cognizant itself had exited the content review business.

In late 2018, I was aware we had a writer who was coming to interview content moderators, and he was given full access. I don't think the executives at Cognizant realized what an unflattering portrait Newton would paint of the company.

His first article, published on February 25, 2019, with the title "The Trauma Floor – The Secret Lives of Facebook Moderators in America," and published in "The Verge," struck with the force of a nuclear bomb. This is how it opened, detailing the problems of Chloe*:

---

* A pseudonym.

The panic attacks started after Chloe watched a man die . . .

For this portion of her education, Chloe will have to moderate a Facebook post in front of her fellow trainees. When it's her turn, she walks to the front of the room, where a monitor displays a video that has been posted to the world's largest social network. None of the trainees have seen it before, Chloe included. She presses play.

The video depicts a man being murdered. Someone is stabbing him, dozens of times, while he screams and begs for his life. Chloe's job is to tell the room whether this post should be removed. She knows that Section 13 of the Facebook community standards prohibits videos that show the murder of one or more people. When Chloe explains this to the class, she hears her voice shaking.

Returning to her seat, Chloe feels an overpowering urge to sob. Another trainee has gone up to review the next post, but Chloe cannot concentrate. She leaves the room and begins to cry so hard she has trouble breathing.[1]

I'm not sure why I didn't develop such severe problems from the images I saw. I certainly was presented with examples just as bad. The first few months were difficult, as sometimes I was affected emotionally by what I saw. But it didn't cause me to have sleepless nights or nightmares. From my perspective, I thought they prepared us well for what we might encounter. Maybe it was my strong religious background that always accepted the idea that there was good and evil in the world, so I wasn't so shocked when I was presented with such examples of depravity.

There were a lot of facts the writer got correct, such as the excessive secrecy surrounding the identity of our employer. He also correctly questions who was really being served by this policy:

Over the past three months, I interviewed a dozen current and former employees of Cognizant in Phoenix. All had signed non-disclosure agreements with Cognizant in which they pledged not to discuss their work for Facebook, or even acknowledge that Facebook is Cognizant's client. The shroud of secrecy is meant to protect employees from users who may be angry about a content moderation decision and seek to resolve it with a known Facebook contractor. The NDAs are also meant to prevent contractors from sharing Facebook users' personal information with the outside world, at a time of intense scrutiny over data privacy issues.

But the secrecy also insulates Cognizant and Facebook from criticism about their working conditions, moderators told me. They are pressured not to discuss the emotional toll that their job takes on them, even with loved ones, leading to increased feelings of isolation and anxiety.[2]

The situation described by Newton also explains why I didn't feel much guilt over breaking my nondisclosure agreement. Since I came forward, I've had the opportunity to talk to several lawyers, and they've told me nondisclosure agreements are rarely upheld by courts. Companies and their lawyers know this but still make their employees sign them, so that if a situation comes up, they can try to bully them into remaining silent.

However, I did take issue with the picture painted of the Cognizant workplace. While I certainly felt pressure to perform, for the most part I felt supported and believed the managers were genuinely concerned about our emotional wellbeing:

> Collectively, the employees described a workplace that is perpetually teetering on the brink of chaos. It is an environment where workers cope by telling dark jokes about committing suicide, then smoke weed during breaks to numb their emotions. It's a place where employees can be fired for making just a few errors a week, and where those who remain live in fear of the former colleagues who return seeking vengeance.
>
> It's a place where, in stark contrast to the perks lavished on Facebook employees, team leaders micromanage content moderators' every bathroom break and prayer break; where employees, desperate for a dopamine rush amid the misery, have been found having sex inside stairwells and a room reserved for lactating mothers; where people develop severe anxiety while still in training, and continue to struggle with trauma symptoms long after they leave; and where the counseling that Cognizant offers them ends the moment they quit, or are simply let go.[3]

Maybe it's just me, but I didn't see any of those extreme examples of what the writer described. I'm not saying such incidents didn't take place. But eventually Cognizant would swell to more than a thousand employees on three floors, working three different eight-hour shifts, (because social media never sleeps!), and so there were bound to be a lot of stories. I made abundant use of the counseling services when I was at Cognizant and even felt I struck up something of a friendship with the lead psychiatrist. After the Casey Newton article came out and slammed Cognizant's counseling department so hard, I went directly to the lead psychiatrist and told him I thought they were doing a great job in trying to protect our emotional wellbeing. I had always felt supported by them. He thanked me, then shrugged his shoulders as if to say, "Writers! They always want to focus on the bad things!"

And if moderating social media content is supposed to bring about more rational discussion on the Internet, it doesn't even seem to be working with the content moderators:

> The moderators told me it's a place where the conspiracy videos and memes that they see each day gradually lead them to embrace fringe views. One auditor walks the floor promoting the idea that the Earth is flat. A former employee told me that he has begun to question certain aspects of the holocaust. Another former employee, who told me he has mapped every escape route out of his house and sleeps with a gun at his side, said, "I no longer believe 9/11 was a terrorist attack."[4]

I want to make it very clear that bad policies by those in charge—and I lay this at the feet of Mark Zuckerberg, the head of Facebook—can have disastrous consequences on employees who are just trying to do the job they've been given to do. In my opinion, the central flaw in Cognizant's content moderation policy came down to one inescapable fact. They wanted content moderation, but they didn't want to pay for a quality job. Who do you get for these jobs when you pay them fifteen dollars an hour?

You get kids who recently graduated from high school, twenty-somethings, and maybe some people like me in their thirties who are thinking it's a good chance to get in on the ground floor of something that might end up being great. The salary should be two or three times what they pay, a college degree should be required at a minimum, and probably a preference for veterans, since, from their service, they'd probably already seen a lot of crazy stuff. Your employees shouldn't be sheltered suburban kids.

Secrets get leaked out. That's just the way human beings are wired. And it wasn't Casey Newton, or even I who first started raising questions about Facebook's content moderation policy, but *The Guardian* newspaper from England in May 2017.[5] Newton detailed this history in his article:

> Last April, a year after many of the documents had been published in the Guardian, Facebook made public the community standards by which it attempts to govern its 2.3 billion monthly users. In the months afterward, Motherboard and Radiolab published detailed investigations into the challenges of moderating such a vast amount of speech.

Those challenges include the sheer volume of posts; the need to train a global army of low-paid workers to consistently apply a single set of rules; near daily changes and clarifications to those rules; a lack of cultural or political context on the part of the moderators; missing content in posts that make their meaning ambiguous; and frequent disagreements among moderators about whether the rules should apply in individual cases.

Despite the high degree of difficulty in applying such a policy, Facebook has instructed Cognizant and its other contractors to emphasize a metric called "accuracy" over all else.[6]

Bad systems lead to bad results. In light of what happened, the article got many things right. They were trying to buy content moderation on the cheap. And I can't help but also think that they were trying to curtail free speech in a way that was more beneficial for their political leanings. Sometimes it's only when you leave a situation that you understand how truly crazy it was. This is how the article described the experience of one employee, Miguel:[†]

When Miguel has a question, he raises his hand and a "subject matter expert" (SME), a contractor expected to have more comprehensive knowledge of Facebook's policies, who makes $1 more an hour that Miguel does, will walk over and assist him. This will cost Miguel time, though, and while he does not have a quota of posts to review, managers monitor his productivity, and ask him to explain himself when the numbers slip into the 200s.

From Miguel's 1,500 or so weekly decisions, Facebook will randomly select 50 or 60 to audit. These posts will be reviewed by a second Cognizant employee, a quality assurance worker, known internally as QA, who also makes $1 per hour more than Miguel. Full-time Facebook employees then audit a subset of QA decisions and from these collective deliberations, an accuracy score is generated.[7]

In some ways, I feel a sense of pride in lasting at Cognizant until the end. My job had been crazy for just shy of two years, but in November 2019, the decision was made that they were exiting the content review business, ending things by February 2020. Again, Casey Newton reported the news:

Yesterday, as I tried to sort through Twitter's decision to ban political ads, I got a tantalizing tip from a new source. Cognizant, the professional services

---

† A pseudonym.

company I have spent much of the year investigating over the dire conditions of its workplaces, was exiting the content moderation business . . .

Cognizant intends to finish their contracts, which will begin to rap up March 1st and then wind down throughout the remainder of 2020. Both of the sites I visited are closing as a result of Cognizant's announcement yesterday, affecting more than 6,000 employees around the world . . .

A memo from CEO Brian Humphries to all employees that someone sent me let them know that, while thousands of jobs would be eliminated, Cognizant would make a donation intended to spur the development of machine-learning systems that can take the place of human moderators.[8]

The article also went on to note that the closure of Cognizant, in addition to affecting Facebook, would also affect Google and Twitter, which also had contracts with Cognizant, but neither company revealed the size of those contracts.[9]

And of course, there was the inevitable legal settlement Cognizant had to pay to moderators who were traumatized by what they'd experienced. In May 2020, shortly before my video appeared on Project Veritas, it was reported:

In a landmark acknowledgment of the toll that content moderation takes on its workforce, Facebook has agreed to pay $52 million to current and former moderators to compensate them for mental health issues developed on the job. In a preliminary settlement filed on Friday in San Mateo Superior Court, the social network agreed to pay damages to American moderators and provide more counseling to them while they work.

Each moderator will receive a minimum of $1,000 and will be eligible for additional compensation if they are diagnosed with post-traumatic stress disorder or related conditions. The settlement covers 11,250 moderators, and lawyers in the case believe that as many as half of them may be eligible for extra pay related to mental health issues associated with their time working for Facebook, including depression and addiction.[10]

While I could be happy in some ways that there was finally some acknowledgment of how terribly Cognizant had run its business, on the other hand it was completely maddening. Moderators got an extra thousand dollars, in addition to possibly some counseling, and that was supposed to make it all better? Recall that the average Facebook employee makes $240,000 a year. The disparity in pay was shocking.

I was frustrated by another part of this news. It all seemed like a bit of clever misdirection, the way a magician will get you to look at one hand, while you don't see what he's doing with the other. Yes, content moderation took a terrible toll on those unprepared employees whom Cognizant hired. The overall message was that people in the world could be terrible, and I couldn't agree more with that opinion. There are some genuinely evil people out in the world. However, by focusing on that message, the question of what Facebook itself was doing was obscured. It's the old game played by every tyrant in the world. They convince the public there's a terrible threat, then hope the public won't look too closely at what the government is doing to those people who are identified as a "threat." Maybe those people identified as "threats" are just the peaceful, law-abiding opposition, not bomb-throwing terrorists.

Cognizant might have fallen, but the challenge posed by "content moderation," possibly by machines using preprogrammed algorithms, still existed. The tyrants hadn't given up the game of trying to control thought.

They were just trying to build new, and what they hoped would be better, weapons, without the fragile flaws of human beings.

* * *

With the release of the Project Veritas video, I became a worldwide celebrity.

Since I'd spent most of my time at Cognizant working on Latin America, it's probably not surprising that I got a lot of attention from conservative politicians in Brazil, the largest democracy in South America. The current president of Brazil is Jair Bolsonaro, elected in 2018, and he's been described as the Latin American Donald Trump. Like Trump, Bolsonaro had many supporters in their Congress but also faced significant opposition from the more liberal forces in his country, such as their Supreme Court.

The reason Brazil took notice of me is because of a few paragraphs I wrote as part of a joint letter to Congress written by myself; Zach Vorhies, the Google Whistleblower; and Zachary McElroy, the other Facebook whistleblower.[11] We wrote the joint letter in July 2020, and in my portion of the letter I say the following with regard to Brazil:

> Over the last month I've had the opportunity to do media interviews with over 20 different news outlets, both in the U.S. and abroad. In my interviews in Spain, Canada, and Colombia, I noticed that citizens are concerned about Facebook's reach and influence in their elections. Conservatives in these

countries also suffer political censorship. This is not just a domestic concern, this is a global pandemic; political censorship has infected the world in a major way, including in Brazil where the federal judiciary is colluding with Facebook and Twitter against Brazilian President Jair Bolsonaro. Facebook and Twitter are quick to fold when strongarmed by a corrupt federal mob (equivalent of mainstream media in the United States) purely for political purposes. Sad, indeed.[12]

In large part because of this letter, I caught the attention of Brazilian journalist Allan Dos Santos and did a livestream with him and Congresswoman Bia Kicis. During this interview we talked about tech censorship and how Brazilian conservatives were actively being censored. Subsequently, I was invited specifically by Congresswoman Bia Kicis, an outspoken supporter of Bolsonaro, to visit Brazil and share my message with as many groups as possible. The only real drawback was that, despite my newfound celebrity, I was still kind of broke. Even at Cognizant I'd only been making twenty-eight thousand a year, and the plan was for me to fly into Brazil on September 2, 2020, and stay until the 11th. It was a great plan, but I didn't have the funds to make it happen. Keep in mind that I had resigned my full-time job in June 2020 working for a mortgage repair call center and never received any unemployment paychecks.

In late August, I decided to do a live stream and put out an appeal for money. I was fortunate that one listener gave me her frequent flyer miles, and I raised a couple thousand dollars to get me through my time in Brazil and pay for rent at home. Another complication was that several weeks earlier I'd put a rush on my passport and wasn't sure whether it would arrive in time for my flight. Even though I hadn't yet received my passport, with the frequent flier miles in hand I booked a flight from Phoenix to São Paulo, Brazil. I'd mentioned to my lovely wife, Livy, that I was "thinking" about going to Brazil, but when I said to her, "Hey, I'm leaving next week and I'll be gone for nine days," she was a little upset. Yes, life around me can be a little unpredictable, but she eventually forgave me. My passport arrived in time, so I was off to Brazil.

I was picked up in São Paulo by Fernando Lisboa, who has multiple YouTube channels in Brazil with a total of more than five million subscribers. His primary YouTube channel, *Vlog do Lisboa*, has nearly a million subscribers alone. In the United States, I'd compare him to political commentator Steven Crowder, whose show *Louder with Crowder* is known for its irreverence and sharp, political commentary. Lisboa filled me in on the situation in Brazil, with which I'd had some familiarity. Apparently, the Brazilian Supreme Court was flirting with a "Fake News" claim against certain conservative journalists in order to stop their broadcasts. Three months earlier, the Brazilian Supreme Court had apparently authorized the seizure of cell phones and laptops belonging to two conservative commentators for the alleged crime of spreading "fake news," which actually meant showing support for Bolsonaro. I met with both of these individuals. Marcos Bellizia sits on the board of Boston University and was targeted by the Supreme Court. Otavio Fakhoury is a businessman and former Wall Street investor who contributed financially to a right-wing news outlet, and as a result his home was raided and his property seized illegally.

In addition to this, the Brazilian Supreme Court had issued an order, essentially asking Facebook and Twitter to ban twenty-nine political figures from Brazil. This ban would restrict these individuals from using these social media platforms globally.

While I was in Brazil, my interviews were scheduled by an activist attorney named Flavia Ferronatto. During various meetings throughout my time in Brazil, we discussed the idea of using the Global Magnitsky Act as a way to respond to the Brazilian Supreme Court's brazen attempts to silence conservative activists. The Global Magnistky Act is a tool that can be used by the president of the United States to place sanctions on human rights abusers in foreign countries.[13] We thought it important that since the Brazilian Supreme Court was taking actions against conservatives with American ties, namely, Marcos Bellizia and Otavio Fakhoury, perhaps the United States could intervene and help Brazil fight against the totalitarian control of the Brazilian Supreme Court.

On one of my first days in Brazil, I was chauffeured around in a bullet-proof Mercedes by another wealthy, conservative Brazilian, Fabrizio Fasano, Jr. He talked with me extensively about my experience working with Facebook, while he showed me the various hotels his family owned and took me for lunch at one of their restaurants. Fabrizio is somewhat of a celebrity in Brazil and once hosted a popular cooking show on national television. However, once he went public with his support for President Bolsonaro, he received severe backlash. Despite the pushback, Fabrizio proudly and publicly supports the conservative ideals espoused by those of the political right in Brazil, because he sees the left as being co-opted by communism and wants a better future for his children. He now has another television program called *Programa Coliseum*, and while I was in Brazil we filmed an interview together on his show.[14]

On Monday, September 7, four days after I'd landed in Brazil, Flavia and I took a flight to Brasilia, the capital of Brazil, to celebrate Brazilian Independence Day. The reach of Project Veritas into Brazil was an enormous surprise to me, as I was walking around the celebration with many well-known conservative figures in Brazil, and I was the one who got noticed. Many people came up to shake my hand and thank me for coming to their country. In my nine days in Brazil (really just seven when you consider the flight time), I did seventeen interviews, met with many human rights activists, and even consulted with several of their congressmen.

While in the capital, I visited the studio of *Terca Livre*, one of the largest conservative television channels in Brazil. I also met with Congresswoman Bia Kicis several times and discussed with her Brazil's censorship struggles against Big Tech.

It was clear to me after my Brazil trip that the fight for freedom on the Internet isn't just confined to the United States, but is a worldwide fight.

* * *

When you become a whistleblower, you enter into a small and exclusive club of individuals. I've mentioned the other Facebook whistleblower, Zach McElroy, but after I went public I struck up a friendship with Zach Vorhies, who'd blown the whistle on Google.

Zach had called me up over the summer and asked if I'd be willing to speak at the American Priorities Conference (AMPFest) in October 2020 at the Trump Doral Resort in Florida, with him and Zach McElroy, about a month before the presidential election. I agreed, but after my Brazil trip I genuinely hoped there'd be some way I could gracefully bow out.

When I complained to Zach that I didn't have the money for the trip, he found a way for me to get there and told me I'd have my room at Trump Doral comped. There was no excuse for me not to go, so again I told my lovely wife that I'd soon be off on another adventure. She was more understanding this time. Or at least she did a better job of pretending.

Originally, the three of us were supposed to have one hour of time at the conference, but that got chopped down to thirty minutes. I'd prepared a presentation with seventy slides, and when I arrived, the first thing Zach did was bring me up to his room so we could start cutting things so I could fit into my new seven-minute time slot. Zach is a master showman, and after glancing through my presentation, he said, "Remember, this isn't about you and all the things you know. This is about the audience. What is the one thing you want to start with that will hit them with the greatest force?"

I thought for a moment and then said, "The picture that Facebook allows on its platform of Donald Trump having his throat slit. They say they're against violent images, but that one they allow."

"Perfect," he answered. "Yes. Because it's not Donald Trump that Facebook is targeting. It's every single Trump voter. That's what the audience will take away!"

We worked in Zach's room until about midnight. The air conditioner in Zach's room was out, and as an Arizona boy I can handle heat, but not the humidity. There were a couple times I felt like I almost passed out. Later, I wondered why we didn't go to my room, where the air conditioner was working.

I thought I'd be able to sleep that night, but I must have been too wired. I think I caught a few hours of sleep, but it didn't seem like much.

Our presentation began around one in the afternoon, and I was nervous waiting back-stage. Zach had this special lighted mask, which displayed a waving American flag with near-video quality. He walked onto the stage, took off his mask, and said, "My name is Zach and I'm from the future. And in the future, the communist cabal has been defeated, and Donald Trump is the president of the United States. But now, in 2020, we have our work cut out for us." The audience ate it up, and within a few minutes he introduced me.

I started with the image Facebook allowed on its platform of Donald Trump having his throat slit, ran through some of the more glaring contradictions between what liberals and conservatives were allowed to say, and then handed it off to the other Facebook whistleblower, Zach McElroy.

With my presentation successfully out of the way, I was able to relax and get to know some of the other guests. I talked with George Papadopoulos, the former Trump volunteer who'd been so viciously attacked by Robert Mueller's goons; and also with the brilliant Matt Couch, who has done such a great job revealing unknown facts about the case of Seth Rich, the computer specialist for the Democratic National Committee who died under mysterious circumstances in 2016 and was believed to be the source for Julian Assange and his revelations about the Hillary Clinton campaign.

Before I left, I watched one final presentation on Sunday that blew my mind. It was delivered by Nick Noe and Charles Wood, the father of Ty Woods, who was one of the four military contractors killed in Benghazi, Libya, on September 11, 2011. The two of them presented a video interview with a guy named Alan Parrot, an American of Sikh descent who'd worked in the Middle East for several decades as a falconer. According to Parrot, falconry is the single sport that unites the wealthy of the Middle East with Al Qaeda, and you can't understand how these two groups mutually support each other without understanding falconry.

The story told by Parrot is that Bin Laden was given sanctuary in Iran for several years after the September 11, 2001, terror attacks in New York and Washington, DC, but that the United States and Iran were close to a deal to resolve the problem. Apparently, Iran had agreed that Bin Laden would be held in a neutral third country (because supposedly it's against the Muslim religion for any Muslim to turn over a fellow Muslim to an infidel), but that Obama wasn't satisfied with the deal. Instead, he wanted a "trophy kill" of Bin Laden, which the Iranians agreed to but never planned to follow through with. Instead, our forces supposedly killed a Bin Laden double, which the Iranians used as blackmail against Obama, in exchange for the $150 billion so-called "Iranian deal."

On Saturday, after my talk and before theirs the next day, I had the chance to sit with Nick Noe and Charles Wood, and they seemed extremely nervous that something would happen to them. At one point, Charles Wood said, "Yeah, my wife packed me a bunch of peanut butter and jelly sandwiches because she's afraid I'm going to get poisoned." The story about Bin Laden also implicated Hillary Clinton, as she was secretary of state at the time, and there were allegations that Libya was being used as an arms shipment point for the conflict in Syria.

There has been some coverage of this story by journalists like Anna Khait and Maryam Henein, but many people are hoping that further information will be released to either corroborate or disprove the allegations.

* * *

On July 27, 2020, as a direct result of my coming out as a whistleblower against Facebook, Congressman Matt Gaetz filed a criminal referral with the United States Justice Department against Mark Zuckerberg for false testimony given before Congress on April 10 and April 11, 2018.[15] The letter is reproduced below in its entirety:

The Honorable William Barr

Attorney General of the United States

United States Department of Justice

950 Pennsylvania Avenue, NW
Washington, D.C. 20530

Dear Attorney General Barr:

I write to urge you to investigate the conduct of Mark Zuckerberg, CEO of Facebook, Inc., before the United States Congress.

On April 10, 2018, Mr. Zuckerberg testified in a joint hearing of the Senate Judiciary Committee and the Senate Commerce, Science, and Transportation Committee. The next day, Mr. Zuckerberg testified before the House Energy and Commerce Committee. On both occasions, members of Congress asked Mr. Zuckerberg about allegations that Facebook censored and suppressed content supportive of President Donald Trump and other conservatives. In his responses, Mr. Zuckerberg repeatedly and categorically denied any bias against conservative speech, persons, policies, or politics. Mr. Zuckerberg also dismissed the suggestion that Facebook exercises any form of editorial manipulation. However, recent reports (**https://projectveritas.com/news/ facebook-content-moderator-if-someone-is-wearing-a-maga-hat-i-am- going-to/**) from Project Veritas, featuring whistleblowers who worked as Facebook's "content moderators," have shown ample evidence of such bias and manipulation.

Two content moderators, Zach McElroy and Ryan Hartwig, both worked on the Facebook content review flow generated by Facebook's artificial intelligence (AI) program for flagging questionable content. McElroy worked at the Facebook-Cognizant facility in Tampa, Florida and Hartwig worked at the Facebook-Cognizant facility in Phoenix, Arizona.

On June 23, 2020, Project Veritas published the results of an undercover investigation featuring the aforementioned whistleblowers. Their report revealed that the overwhelming majority of content filtered by Facebook's AI program was content in support of Donald Trump, Republican candidates for office, or conservatism in general. This alone is already an indication of bias within the platform.

Once flagged by Facebook's AI, moderators reviewed the filtered content, and adjudicated whether it qualified as removable. According to the Veritas report and undercover footage, the adjudicators were outspoken about their political bias against Republicans, and actively chose to eliminate otherwise-allowable content from the platform and from public view simply due to its political orientation. This arbitrary and capricious behavior is not done in good faith, and falls outside of the express intent of Section 230 of the Communications Decency Act, which affords Facebook liability protection as long as the platform moderates content in "good faith."

Additionally, these facts are in direct contrast to Mr. Zuckerberg's testimony before Congress where he stated under oath that Facebook is a politically-neutral platform, and that he personally is working to root out any employees who are restricting speech based on Silicon Valley's overwhelmingly leftist cultures. (**https://www.washingtonpost.com/news/the-switch/wp/2018/04/ 10transcript-of-mark-zuckerbergs-senate-hearing/**).

Project Veritas' undercover footage shows that a great deal of "political speech" supporting the President was labeled "hate speech," or was considered in violation of Facebook's "**Community Standards**" (**https://www. facebook.com/communitystandards/objectionable_content**). At the same time, speech promoting violence against the President and his supporters was labeled as merely "political," and was thus allowed to stay on the platform. For example, McElroy captured a shot of a Facebook corporate ruling that an illustration of a hand holding a knife, slashing the throat of the President, captioned by "Fuck Trump," would be allowed as political speech, despite being in clear violation of Facebook's guidelines. In this case, the guidance to content moderators instructed them to watch for hostility directed at the gallery that posted the image.

Facebook's AI screening content is not politically neutral. Neither are the moderators hired to review content flagged by the AI program. This stands in opposition to Mr. Zuckerberg's congressional testimony, and violates the "good faith" provision of Section 230(c)(2)(A) of the Communications Decency Act.

Accordingly, I respectfully refer Mr. Zuckerberg to the Department for an investigation of potential violations of 18 U.S.C., sections 1001, 1505, and 1621 for materially false statements made to Congress while testifying under oath.

Oversight is an essential part of Congress' constitutional authority. Customarily, Congress is grateful to citizens who come forward with relevant information in good faith, as the aforementioned whistleblowers have done. As a member of this body, I question Mr. Zuckerberg's veracity, and challenge his willingness to cooperate with our oversight authority, diverting congressional resources during time-sensitive investigations, and materially impeding our work. Such misrepresentations are not only unfair, they are potentially illegal and fraudulent.

I hope you will give this referral full and proper consideration. If you need further clarification, please contact my chief of staff, Jillian Lane-Wyatt.

Sincerely,
Matt Gaetz
Member of Congress.[16]

When that letter was publicly released, it had a profound impact on me. An ordinary citizen had been listened to by a prominent congressman, and a criminal referral had been made against Mark Zuckerberg, one of the wealthiest people on the planet, for lying to Congress. Is there anything more American than believing the truth will eventually prevail?

But I don't want to get ahead of myself. I know how much Big Tech donates to Congress and that Gaetz's criminal referral will likely not be acted upon, especially now with President Biden in office.

However, every revolution has a starting point.

And it begins with telling the truth.

My coauthor, Kent Heckenlively, told me that when he was in college he wrote a political column for the school newspaper supporting President Reagan's crusade against communism in the Soviet Union. As Kent tells it, his professors thought he was crazy. And yet to Kent it appeared so simple. Communism was crushing the soul of humanity, and it was not something the citizens of those countries wanted. People want to speak freely without fear of reprisal, to worship as they choose, and to live their lives according to the dictates of their own conscience.

Reagan told the truth about this reality. For speaking the truth, he was savaged in the media as being a warmonger.

But he was a champion of freedom. I genuinely believe there is no power in the world stronger than telling the truth. We just need to have the courage to do it, no matter how much our voice may shake in fear as we confront the powerful who want to perpetuate their lies.

When the Soviet Union dissolved on December 25, 1991, the mainstream media was flabbergasted. They'd expected it to last forever, a system with which the West would have to learn to coexist. But as Eastern Europe and Russia were free of this scourge, the question was raised as to why it had taken so long to fall. In later discussion with former communist officials, when asked this question they said essentially, "We thought we were the good guys." But then Reagan came along, and unlike the Western press, or their own *Pravda* newspaper, they said that they were the bad guys. And these communist officials had to consider the question "Are we the bad guys?"

I believe something similar needs to happen in Big Tech and among those on the left.

They really are convinced they're the "good guys," and we're the "bad guys."

It's important to dispel that narrative and change their hearts to see us as people who simply have a different point of view. I choose to do it with my advocacy, this book, and I've also started a foundation, the Hartwig Foundation for Free Speech, to carry on this work: www.ryanhartwig.org/the-hartwig-foundation-for-free-speech/. I hope you will support my efforts.

When I started at Cognizant, I believed in the mission of trying to keep social media as a warm and inviting place where people were free to discuss ideas. I'm sure you realize by now I like to share my views in a respectful manner. That is my personal view, but I don't know how respect can be guaranteed by law.

There are a lot of people, in addition to Congressman Gaetz, who are talking about Section 230 of the Communications Decency Act. One of them is Supreme Court Justice Clarence Thomas, currently the longest serving justice on the court. In turning down a recent case on October 13, 2020, Thomas suggested that the time had probably come for the Court to weigh in on Section 230, which it has never done before. He wrote:

> First, Section 230(c)(1) indicates that an Internet provider does not become
> the publisher of a piece of third-party content, and thus subject to strict lia-
> bility, simply by hosting or distributing that content. Second, Section 230(c)
> (2)(a) provides an additional degree of immunity when companies take down
> or restrict access to objectionable content, so long as the company acts in good
> faith. In short, the statute suggests that if a company unknowingly leaves up
> illegal third-party content, it is protected from publisher liability by Section
> 230(c)(1); and if it takes down certain third-party content in good faith, it is
> protected by Section 230(c)(2)(A).[17]

The important phrase in the opinion of Justice Thomas is likely to be the meaning of "good faith." Has Facebook been operating in "good faith"? That really gets to the heart of the matter, doesn't it?

"Good faith" is one of those things that's hard to define. Is "good faith" created by laws, regulations, or training manuals? I don't think it is. I believe "good faith" comes from a deeper place, the heart, or maybe the soul if you're a religious person. It can't be seen, but it can be felt. In the beginning, I believe Cognizant understood the challenge it was facing, which is why one of the first principles they taught us was "DKR," which stood for "Dignity, Kindness, and Respect." It was how we were supposed to treat one another. But it was a principle they violated first with their customers, then increasingly with their employees. Is it any wonder that the entire venture eventually fell apart?

I find it difficult to live in this world without my faith. Not necessarily because I believe every teaching of the Mormon Church. But because it gives me the framework by which I can be a good person. Do you know why the Mormon Church has the largest genealogical library on the planet? It's because we want to know the names of these people and bless them, so they may enter the Kingdom of Heaven. I understand that through one lens that might appear crazy. But through another lens it's one of the ultimate acts of kindness. My church believes that even if you never met a Mormon, or died before our religion was established, we will take the time to learn your name, stand up for you, and bless you before God.

And I understand that many people in the world don't act that way, especially with the anonymity that social media can provide. People can be terrible to one another online in ways that they would never be in a face-to-face interaction with that same person.

In the presence of such anger and bias, we must be calm and resolute in telling the truth, holding a mirror up to these actions, and asking our opponents if they like the image that is reflected to them.

I think they will recoil from what they see, and perhaps they can begin to understand how they have been unjust.

I'm sure this book will be controversial, and I'll endure vicious attacks and outright falsehoods about my beliefs. But let this be my final message. Treat one another with "good faith." Every person you come across is a unique creation of God. Many are hurt or damaged by what they have experienced in life. Be slow to anger and quick to forgive. Sometimes the most difficult lessons we learn as adults are those we were first taught as children. Think of your first days of school, that kind teacher who loved all of you and taught you how to live together as a community. Many of us may have forgotten these lessons, but they're not difficult to relearn. Think of the joy you experienced when you made your first friend, what you learned about the life of another person, and how the differences of those you met enriched your life. We are not so different as adults. We need to remember how good it feels to connect with other human beings. Be curious instead of judgmental.

And if you're ever in doubt about how to act, just remember to treat everybody you meet with "Dignity, Kindness, and Respect."

If you find yourself categorizing those you treat with good behavior and those you can treat poorly, you might want to consider that you've become the "bad guy."

EPILOGUE

Election Day Problems and Final Thoughts

On November 3, 2020, like many Americans, I cast my vote in the presidential election between President Donald Trump and former Vice-President Joe Biden.

However, after casting my vote, I went to a local polling location to do some campaigning for Merissa Hamilton, who was running for mayor of Phoenix. The mayoral race in Phoenix was nonpartisan, although ideological preferences were clear. I planned on following all rules, and did, but must note some irregularities I observed that day.

The campaign rules did not allow for campaigning within seventy-five feet of the entrance to the polling location, and this was marked by a sign. I set up a lawn chair along with some campaign literature just beyond the sign. This was an ideal location because it gave me access to people coming from several different directions.

However, after I'd been there for about an hour, a poll worker came out and asked, "You're not a Republican, are you? You're not some mole, are you?"

"I'm more of a libertarian," I replied.

"That means you're a Republican," she answered, then turned on her heels and walked away.

A few minutes later another poll worker came out, saying she had to remeasure the seventy-five-foot distance from the entrance to the polling location. She moved the sign twenty feet farther away, which made it more difficult for me to talk to people. However, I dutifully moved my chair.

A third poll worker came out about twenty minutes later. I asked her why the sign had been moved twenty feet and if I could take a picture of her with my phone.

She went off on me and said, "No, you can't take my photo. That's against the law. I'll call the police on you. I'll take your phone away from you."

"You can try and see what happens," I replied and snapped her photo.

She stepped toward me as if to try and grab my phone, and I took a step back. "Whoa! Hold on there," I said.

She turned and walked off in a huff.

I stayed about another half hour, then left to run some errands for my wife.

As the elections results trickled in that day and into the next, a feeling arose among many people that something was wrong with the election counting. Arizona was the center of some of the controversy, with allegations in Maricopa County that poll workers had handed out Sharpie pens to those thought to be Trump voters, which the machines would invalidate. I spoke at a couple of rallies, given my newfound celebrity as the "Facebook whistleblower," and on the Saturday night after the election was tracking some vehicles and an airplane we were told contained phony ballots to be counted. I filled out an affidavit about what I saw and cannot say any more about that incident at the time of this writing.

In the week after the election, I was also doing a lot of interviews with Brazilian outlets about what was happening in the United States.

And it was a good question.

What did happen in the election of 2020?

* * *

Sometimes you need to just slow things down and take it piece by piece. Of all the articles and accounts I've read, I think the three reports put together by White House Director of Trade and Manufacturing Policy for Trump Peter Navarro laid out the clearest case for why so many people are angry about what happened on that day, and those that followed. This is from the introduction to his first report:

At the stroke of midnight on Election Day, President Donald J. Trump appeared well on his way to winning a second term. He was already a lock to win both Florida and Ohio; and no Republican has ever won a presidential election without winning Ohio while only two Democrats have won the presidency without winning Florida.

At the same time, the Trump-Pence ticket had substantial and seemingly insurmountable leads in Georgia, Pennsylvania, Michigan, and Wisconsin. If these leads held, these four battleground states would propel President Trump to a decisive 294 to 244 victory in the Electoral College.

Shortly after midnight, however, as a flood of mail-in and absentee ballots began entering the count, the Trump red tide of victory began turning Joe Biden blue. As these mail-in and absentee ballots were tabulated, the President's large leads in Georgia, Pennsylvania, Michigan, and Wisconsin simply vanished into thin Biden leads.

At midnight on the evening of November 3, and as illustrated in Table 1, President Trump was ahead by more than 110,000 votes in Wisconsin and more than 290,000 votes in Michigan. In Georgia, his lead was a whopping 356,945; and he led in Pennsylvania by more than half a million votes. By December 7, however, these wide Trump leads would turn into razor thin Biden leads—11,799 votes in Georgia, 20,682 votes in Wisconsin, 81,660 votes in Pennsylvania, and 154,188 votes in Michigan.[1]

Let's consider the craziness, shall we? Trump had already won Ohio and Florida, something Republican candidates had to do. It was down to four battleground states, Wisconsin, Michigan, Georgia, and Pennsylvania.

At midnight Eastern time on November 3, Trump was leading Biden in these states by 1,317,208 votes.

On December 7, 2020, the vote tally in these four states had shifted to Biden, who was leading Trump by a little more than a quarter million (268,309) votes.

Got the difference? At midnight on Election Day, Trump was leading in these states by more than a million, three hundred thousand votes. A month later he's losing by a little more than a quarter million votes. That is a swing of more than a million and a half votes in just four states, all from ballots received after midnight. Any fair-minded person should consider these uncontested facts to be deeply troubling. Navarro's report goes onto detail troubling reports out of two other states:

There was an equally interesting story unfolding in Arizona and Nevada. While Joe Biden was ahead in these two additional battleground states—by just over 30,000 votes in Nevada and less than 150,000 in Arizona—internal Trump campaign polls predicted that the President would close these gaps once all the votes were counted. Of course, this never happened.

In the wake of this astonishing reversal of Trump fortune, a national firestorm has erupted over the fairness and integrity of one of the most sacrosanct institutions in America—our Presidential election system. Critics on the Right and within the Republican Party—including President Trump himself—have charged that the election was stolen. They have backed up these damning charges with more than 50 lawsuits, thousands of supporting affidavits and declarations, and seemingly incriminating videos, photos, and first-hand accounts of all kinds of chicanery.[2]

Let's simply say there's good evidence that four states—Wisconsin, Michigan, Pennsylvania, and Georgia—had something extremely odd happen in them. A swing of more than a million and a half votes after midnight on election day is probably unprecedented in American history. If the trends had remained the same after midnight as they were before midnight, then the Electoral College vote would have been 294 to 244 in the Electoral College in favor of Trump, with only 270 needed for a victory.

Trump was behind in Arizona and Nevada, but the Trump campaign internal polls indicated they'd close the combined gap of approximately 180,000 votes. Note that this number is much smaller than the million and a half vote swing in the other four battleground states for Biden. What would the Electoral College numbers have looked like in this scenario? Arizona has eleven electoral votes and Nevada has six, which would add seventeen electoral votes to Trump and take the same amount away from Biden.

In this scenario, Trump would have received 311 Electoral College votes to 227 votes for Biden, which would have been a slight improvement over Trump's 2016 victory over Hillary Clinton of 304 electoral college votes to 227 votes (seven electors in that election declined to state a preference). The introduction concluded:

> That the American public is not buying what the Democrat Party and the anti-Trump media are selling is evident in public opinion polls. For example, according to a recent Rasmussen poll: Sixty-two percent (62%) of Republicans say it is "Very likely the Democrats stole the election" while 28% of Independents and 17% of Democrats share that view.

> If, in fact, compelling evidence comes to light proving the election was indeed stolen after a *fait accompli* Biden inauguration, we as a country run the very real risk that they very center of our great union will not hold.[3]

Navarro published his first report on December 15, 2020, roughly six weeks after the election. Do you think the number of skeptical Republicans, Independents, and Democrats has decreased? I think they've only grown in number. And I share Navarro's concern that we "run the very real risk that the very center of our great union will not hold." If a sizable percentage of the public is convinced that something was not fair, how do you convince them otherwise? Do you simply cancel them, or even worse, label them as "domestic terrorists"? It seems to me that you'd want to aggressively investigate such claims.

In his first report, Navarro goes on to detail the five different claims of election irregularities: outright voter fraud, ballot mishandling, contestable process fouls, voting machine irregularities, and significant statistical abnormalities. Navarro concludes his first report by stating:

> From the findings of this report, it is possible to infer what may well have been a coordinated strategy to effectively stack the election deck against the Trump-Pence ticket. Indeed, the patterns of election irregularities observed in this report are so consistent across the six battleground states that they suggest a coordinated strategy to, if not to steal the election, then to strategically game the election process in such a way as to unfairly tilt the playing field in favor of the Biden-Harris ticket . . .
>
> In refusing to investigate a growing number of legitimate grievances, the anti-Trump media and censoring social media are complicit in shielding the American public from the truth. This is a dangerous game that simultaneously undermines the credibility of the media and the stability of our political system and Republic.[4]

Navarro clearly states there is no single "silver bullet" in showing this election was stolen, but the weight of the evidence suggests a coordinated plan to steal a little from here, and a little from there, until it made a staggering difference.

Navarro released a second report on January 5, 2021, titled "The Art of the Steal," which went into further detail. From the executive summary:

The Democrat Party used a two-pronged Grand "Stuff the Ballot Box" Strategy to flood six key battleground states—Arizona, Georgia, Michigan, Nevada, Pennsylvania, and Wisconsin—with enough illegal absentee and mail-in ballots to turn a decisive Trump victory into a narrow and illegitimate Biden alleged "win."

Prong One dramatically INCREASED the amount of absentee and mail-in ballots in the battleground states. Prong Two dramatically DECREASED the level of scrutiny of such ballots—effectively taking the election "cops" off the beat. This pincer movement resulted in a FLOOD of illegal ballots into the battleground states more than sufficient to tip the scales from a decisive legal win by President Trump to a narrow and illegitimate alleged "victory" by Joe Biden.[5]

This second volume puts a lot more meat on the bones of the allegations of voter fraud and gives an insight into how it could be done. Like an Agatha Christie novel in which there are many who participated in the murder of the victim, but each has a bit of deniability because when looked at in isolation, no single action changed the result. Think of what a truly ingenious plan was devised by the Democrats. Flood the system with illegal ballots and then dramatically decrease the enforcement of election rules. It's like opening the gates of the local prison, then telling the local police to stand down. I encourage you to read Volume 2 to see how abundantly Navarro has documented these claims.

In Volume 3, titled "Yes, President Trump Won," released on January 14, 2021, Navarro goes through the evidence, coming to the conclusion that more than three million illegal votes were likely fed into the system through these various strategies, resulting in a Biden victory margin in these battleground states of a little more than three hundred thousand votes. In other words, the likely illegal votes were more than ten times the amount of Biden's margin of victory. Navarro leaves no doubt as to his opinion about this sad chain of events:

> In light of this evidence, this must also be said: Those American citizens who are now questioning the potential illegality of votes cast in the 2020 election should NOT be subjected by cable news networks, social media platforms, or the print media to the kind of abhorrent behaviors that we are now observing—social and political behaviors that are far more worthy of Communist China authoritarianism than American democracy.

> From public shaming to de-platforming, doxing, and public calls to punish and shun all those who have supported the president or worked in his administration, these types of behaviors are not the American way. Rather, this is Orwell, Kafka, and Xi Jinping all rolled up in the death of the First Amendment and the death knell of our democracy.[6]

It's difficult for me to find a single thing wrong with anything in Navarro's conclusion.

Many went to bed on election night with Trump cruising to an easy victory and awoke to find him losing by a narrow margin because of a swing of a million and a half votes. And leading the charge against Trump were the tech giants.

And what happened to Peter Navarro, the former White House director of Trade and Manufacturing Policy? On January 27, 2021, Navarro appeared on *Tucker Carlson Tonight* to tell the host that he'd been banned by Twitter. This is part of what Navarro said in his interview with Tucker Carlson:

> The point here is it's because of who I am and what I might say, not because of anything I did. And this is Pichai at Google, Dorsey at Twitter, Zuckerberg at Facebook, and Bezos at Amazon. They somehow believe that it's their right to shut up half of America. They're doing violence to the First Amendment. They're doing violence to this country.[7]

Carlson appeared deeply concerned by Navarro's description of events but seemed most confused when Navarro additionally claimed he hadn't made a tweet in weeks. However, little of this was surprising to me. As a content moderator, I thought it was clear the Big Tech companies were not operating in good faith.

And I couldn't help but notice that Navarro correctly identified Mark Zuckerberg at Facebook as a member of the technological cabal that had so deeply damaged the American republic.

* * *

What does this herald for social media platforms like Facebook? Can they continue to enjoy their financial success if they show such contempt for half the country?

Personally, I believe Facebook and other similar platforms have entered a period of unprecedented danger. Social media platforms offer us the opportunity to engage with one another. It's clear many people behave in ways online that they'd never do in person. But as much as the platforms offer us the opportunity to engage with one another, they are increasingly engaged in a conversation with us, by limiting the opinions and perspectives of other voices. Put simply, many do not trust the decisions made by these Big Tech censors. And we suspect these censors have actual contempt for us, which is probably the most dangerous warning sign in any relationship.

Dr. John Gottman studied mathematics at MIT but works as a psychologist and has written extensively about marriage, including publishing a five-hundred-page book called *The Mathematics of Divorce*. He claims to have a high level of success in predicting whether a couple will break up, often based on listening to just a single conversation between the couple. Gottman was profiled in the *Blink* by well-known writer Malcolm Gladwell. As Gladwell explains, Gottman looks for a single factor in the interactions of a couple to determine whether they will divorce.

> If Gottman observes one or both partners in a marriage showing contempt for each other, he considers it the single most important sign that the marriage is in trouble.
>
> "You would think that criticism would be the worst," Gottman says, "because criticism is a global criticism of a person's character. Yet contempt is qualitatively different from criticism. With criticism I might say to my wife, 'You never listen, you are really selfish and insensitive.' Well, she's going to respond defensively to that. That's not very good for our problem solving and interaction. But if I speak from a superior plane, that's far more damaging, and contempt is any statement made from a higher level. A lot of time that's an insult: 'You are a bitch. You're scum.' It's trying to put that person on a lower plane than you. It's hierarchical."
>
> Gottman has found, in fact, that the presence of contempt in a marriage can even predict such things as how many colds a husband or wife gets . . .[8]

Think about the consequences of what Gottman found. Contempt is such a powerful force it can affect the very functioning of your immune system, in addition to being the most accurate indicator of the imminent failure of a relationship. Like an unhappy marriage that may linger for years but suddenly disintegrates over what appears to be the slightest provocation or changed circumstance, Facebook and other media platforms are in similarly perilous waters.

We may not have made our voices yet heard in the mainstream media. But just because something isn't talked about in the highly controlled mainstream media doesn't mean it's mythical. We feel the contempt.

Like a spouse in an unhappy marriage, many of us are looking for the exit. We may not have found it yet, but rest assured many see the opportunity generated by our unhappiness. While Facebook and others are multibillion-dollar giants, it also means they are uniquely vulnerable to sudden downturns in business.

The dominoes can fall very quickly when there is no trust.

In all likelihood, the break is now irrevocable. New training manuals, slide decks, mission statements, or guiding principles are unlikely to restore trust because they do not change the heart. Pretty words cannot cover up the darkness in the soul. There is no light or love in you toward many of your fellow human beings, despite your phony expressions of "dignity, kindness, and respect."

Many have glimpsed what lies behind the mask and do not like the ugliness they have seen. When the end comes, however, we will not regard you with contempt, but with pity.

You were smart but lacked the wisdom to trust the people.

You had the world in your hands but betrayed us all with a Judas kiss.

APPENDIX A

Transcripts That Didn't Make the Final Cut

I filmed with a hidden camera for about nine months total. As you can imagine, that is a tremendous amount of footage. Nearly all of the video I transcribed myself each day after work. From the hundreds of hours of footage, Project Veritas produced a twenty-minute video with some of the most captivating conversations. Because of the time limits, many conversations were left out. I've included the transcripts of some of those conversations here. These interactions will give you a deeper view into the mindset of content moderators, and how they viewed their role as Internet censors. It also gives an unparalleled perspective into how they view Facebook's overarching online presence.

A few days before the project ended, I struck up a conversation with Amy Whiting,* whom I had seen around a lot on the North America side but hadn't interacted with very much. She had previously managed the teams handling IGPR (Instagram Profile Review) and INA (which deals with authenticating accounts).

The following was our conversation. It was a rather short conversation and was toward the end of the day, after I had clocked out but was still on the production floor.

Ryan: See ya, Amy!

---

* A pseudonym.

Amy: Bye, do you have tomorrow off?

[Small talk, her last day is tomorrow]

1:51 **Ryan:** Which team do you manage?

1:55 **Amy:** Yeah, Ex-IGPR and ex-INA.

2:19 **Ryan:** Can you create an Instagram name, like Bernie Sanders?

2:25 **Amy:** On Instagram you don't have to be your authentic self.

2:35 **Ryan:** Cuz in 2016 Shawn was telling me they deleted the word deplorable . . . that's good that they're deleting deplorable, I was wondering if they're doing the same thing this year.

2:50 **Amy:** Instagram has a little more space for that, Facebook, it would have to be a page. If someone reported it. If not, then you're good. I mean until someone reports it.

3:00 **Ryan:** Well isn't Facebook like biased though? Aren't they trying to like—

3:02 **Amy:** Silence Shit.

3:04 **Ryan:** Who are they trying to like silence?

3:04 **Amy:** [mouths something without speaking]

3:09 **Ryan:** Like conservatives?

3:12 **Amy:** I agree, yes.

3:13 **Amy:** [whispers] Conservatives.

3:14 **Amy:** [whispers] Facebook's liberal as fuck.

3:16 **Amy:** They try to even it out, but it's kind of clear. Same with everything. It's kind of funny.[1]

The day prior to my conversation with Amy, I also took advantage of some time immediately after clocking out to talk to someone from the afternoon shift. Angela Krutze† worked the evening shift from about 4:30 p.m. to 1 a.m. I had been hired a short time after her, and she was on the policy team, so she interfaced with Facebook on occasion.

Video 10 Angela Krutze

2020/02/25

2:09 **Ryan:** How's my policy team doing over here. Night shift.

2:13 **Angela:** Good, ready to go home.

2:14 **Ryan:** I know, 3 more days and then that's it.

2:18 [We talk about possible future jobs that we have lined up. She starts her new job on Monday.]

[I ask her about her new job]. What are you doing? Like a quality thing?

2:42 **Ryan:** It's probably with Facebook huh?

2:43 **Angela:** Yes, Instagram, Snapchat. [sarcasm] I heard Snapchat's actually in town [in the Phoenix area], just nobody can figure out where they are. They're just way better at being secret than we are.

2:55 **Ryan:** That's pretty impressive, yeah we didn't do a very good job at it.

2:57 **Angela:** I heard they were here, but I have no idea about it. So, Tik Tok's here too. Yeah we tried to bid for TikTok account we lost it. Yeah so someone else has it in the area. Somewhere, in like Mesa.

3:12 **Ryan:** I'm just sad that I'm not gonna be here to be able to see, like Facebook help, make sure Trump loses, like I'm Republican but I'm like a never-Trumper, so I'm just glad that Facebook's like, trying to make sure he loses. I don't know, cuz they are right? Facebook's trying to make sure . . .

---

† A pseudonym.

3:28 Angela: I don't know. That'd be kind of fucked up though, don't you think?

3:31 Ryan: It would be, I mean, you know better than me because you've been on the policy team, you've seen like more conversations with the client, right?

3:39 Angela: Yeah they don't really, I don't think I talk to them that much. They're cockroaches about that kind of sneaky chasey stuff that they're doing.

3:47 Ryan: But you think they are doing it?

3:48 Angela: I don't know. Um, I think they're more likely to make a right-wing person a hate figure, but it's easier to make them a hate figure because the policies are around things that they normally say being considered hate speech. [emphasis added]

4:05 Ryan: That normally the policy is around things that normally the right wing people say?

4:09 Angela: Yeah like crazy people, like it's built that way. So it's easier to do that. But I wouldn't necessarily see them bending things to make it fit, you know. It was just already built up that way with a different kind of mindset.

4:25 Ryan: I mean it should be that way, because you know a lot of these far-right conspiracy people, they're just, like a lot of them are just Nazis, I mean yeah they should be censoring things like that.

4:40 Angela: Well yeah you don't know the backend and what they have the algorithms doing, and stuff like that, so kind of like [does sneaky keyboard gesture].

4:49 Ryan: But you've never seen that, cuz you've met with the client like maybe a handful of times.

4:53 Angela: Just on video chat, we talk about things . . . I'm sure they have a whole bunch of algorithm stuff going on that we don't even know about.

5:05 Ryan: Cuz they've taken a keen interest in the election.

5:09 Angela: Yeah we have a whole queue for it now. I don't know if you've seen it. Civic harassment queue. It's mostly just people calling each other bitches.

5:18 Ryan: Yeah I think I'm in that queue sometimes.

5:20 Angela: If anyone posts on a public figure that's politically related's account, automatically [unintelligible]. It picks up keywords too, like possibly harassment.

5:35 Ryan: So it'll like scrape the system? Like the AI bots will like scrape the system?

5:40 Angela: Yeah they'll pull everything and then they'll report everything that has a keyword. So it's mostly just like "bitch" and "hoe." And any word that has "kill" in it. Stuff like that. Really simple. I don't think it's that good yet. It's pretty new.

5:52 Ryan: And then you've been on the policy team for what, a couple years now?

5:57 Angela: Uh, at least over a year.

6:00 Ryan: Yeah it's been fun, it's been a fun adventure. I mean there's no other job in the world where you can be sitting with your bros about like hey, is this squeezing a female breast or not?

2020/02/25 END OF INTERACTION[2]

Back in November, I had a lengthy conversation with James Liepzig[‡] and Thomas Conradson.[§] They were friends and sat next to each other. James was very thorough and as a writer of novels had some very in-depth thoughts and views about Facebook. Thomas was very opinionated about his view of Facebook and social media, especially when it came to first amendment speech rights. This conversation provides a great insight into how content moderators viewed their job, which consisted of deleting online speech.

---

‡ A pseudonym.
§ A pseudonym.

11/8/2019 Thomas Conradson & James Liepzig

2:10 **Ryan:** Cuz Facebook's gonna get attacked in congress a lot, cuz it's like well why is there this double standard, like why are you like banning, YouTube probably does it more, like why are you banning this university that pushes right wing stuff, like Praeger University, so how can you prove that they're being biased?

2:35 **Thomas:** "That's the thing, I don't think you can prove that they're biased, but I mean we see every day in content, a lot of the stuff that we take down, **it's like yeah you're expressing your political opinions, but you're doing so in a way that violates our content policies, like, don't violate policy, read up on policy, see what you can and can't say, and you won't have this issue.** [emphasis added]

2:57 **James:** All of this would not be happening if people just said like "some" in their post because "some" means it's not hate speech, it stays up. **But they don't even have the cognitive ability to have that nuance.** [emphasis added]

3:12 **Thomas:** "Like we've had this thing, you know, we had that update saying that calling illegal immigrants criminals is not violating, they're just discussing immigration status.

3:22 **James:** As long as it's explicitly clear that they're talking about crossing the border.

3:25 **Thomas:** But it's like, you can say that! No one is saying you can't say that. Just say that, and then don't say we need to kill them all, that they're all filthy wild animals that need to be put in cages. **Like it's really easy to not get taken down, just don't use hate speech.** And it's like you know, sure, a lot of these guys are like you know "what I said wasn't that bad, what about freedom of speech", well again, you're posting content on platforms that are owned by private companies. [emphasis added]

3:58 **Ryan:** "Yeah, well and that's the whole debate, is whether, yeah."

4:00 **Thomas: We're not infringing on your first-amendment rights because you're on Facebook, you don't have 1st amendment rights, you have Facebook speech rights.** [emphasis added]

4:07 **Ryan:** "But hang on, is Facebook a publisher or a platform? So that's the whole . . ."

4:15 **Thomas:** I mean they do publish some content now, but by and large it's still a [unintelligible].

4:20 **James:** Another, just public thing is a lot of them are all, well, people should be able to refuse service to whoever they want, well, they wouldn't discriminate against gay people, but when we refuse to give them service when they say "advocating violence."

4:38 **Thomas:** Like I have a friend in Seattle, who, you know, right after I started working here, he's pretty conservative, and we got in this argument over it, it's like well yeah you're censoring speech. **Yeah okay, imagine you're, imagine that you go to the Facebook bar, you know, you're sitting down, you know, you and your friends each have a drink out and you're just hanging out, and all of a sudden your friend stands up and starts screaming about you know, racial slurs, you'll get kicked out of the bar. That's just what happens.** It's just how people, how society's, we're trying to conduct ourselves and it's like no. You can have fringe opinions, and no one's gonna come after you, just get taken off the platform. You just get your post deleted. [emphasis added]

5:22 **Ryan:** And it's tough because like the Internet's such a new thing, like our laws are not designed to like deal with the Internet. Internet law is still just really hard to deal with.

5:34 **Thomas:** I mean, you just gotta deal with it as you can. It's like yeah, it's like you know, FBI's not gonna kick down your door for saying what you feel about immigrants or whatever, but we're gonna delete your post.

5:55 **James:** But there are still cases where people make like personal threats where the FBI does show up [unintelligible] said shoot Trump.

6:07 **James:** Another thing about the Congress being down Facebook's ass, like oh, you're using people's data incorrectly, and like why are you mad, you're mad because you have competition in this [unintelligible]. [laughter]

6:22 **Thomas:** You guys are running the NSA. You are literally spying on people. You're not just making profit off it.

6:30 **James:** People are making memes about having FBI agents who watch their phones and their phone calls.

6:35 **Thomas:** Like this is a known thing.

6:45 **Ryan:** But yeah there's a lot of stupid, like right wingers who just, they're basically bots because they're just repeating, like everything they see from the media. Like, each side brainwashes the other side, like let's be honest, each side gets brainwashed.

6:57 **Thomas:** Yeah there's a lot of like Occupy democrat shit or like, you know, again, **there's a definitely a lot of bullshit coming from both sides of this but it's like, [there's a] reason one side gets censored more than the opposite, because they use inflammatory terms.** [emphasis added]

7:17 **James:** Or because their people are like "Justin Trudeau needs to be executed because he boosted my gas prices." [laughs].

7:23 **Thomas:** Right, you know we need to hang Hillary and Barack Obama for I don't know whatever they feel like they did.

7:30 **James:** Treason.

7:30 **Unknown Coworker:** I literally had a job yesterday, where this guy was like upset that his free speech was being limited, and then it was like bro, you posted a screenshot, where you said "all Muslims in America should be executed for treason."

[Nate chuckles][3]

Daniel Will was one of my coworkers who appeared in the Project Veritas video release. He also had some very strong opinions about Facebook and where they stand ideologically. He also views Facebook's motive as trying to make more money and avoid looking bad toward their shareholders. He thinks Facebook is just simply catering to whatever their stakeholders want, and that is leading them to censor voices on the right.

Daniel also had some insights into how Facebook already picked positions, and on why they banned Alex Jones. Here are transcripts from two separate days when we chatted about these topics:

12/21/2019 Daniel Will

2019/12/21 05:43:08 VIDEO 34

3:55 **Ryan:** So Daniel, we keep on talking about the stakeholders, right, who are the stakeholders then?

4:06: **Daniel:** People who own stock in the company, even the CEO . . .

4:33 **Ryan:** Where does the bias come from then?

4:36 **Daniel:** The bias comes from the people themselves. So people across the states, people across other countries, create the bias, and then the bias goes towards the stakeholders and how they decision. So if we're leaning left and we say this is hate speech, this is cruel, Facebook's gonna go, well this is hate speech this is cruel, we have to action this way so we don't upset our community.[4]

12/10/2019 Daniel Will

6:49 **Ryan:** So if it came down to Sanders vs Trump, do you think that Facebook would want, would be gunning for Trump or for Sanders to win then? Would that be their stakeholders, would that be their base, would that protect them more?

7:02 **Daniel:** I don't think there's any, the only thing that we'd be protecting depends on what content comes out during that time. It's so dependent on that, and how it's treated. I mean, they already chose to not, **they chose to be biased by allowing misinformation campaigns, but uh, no, allowing campaign misinformation campaigns, but not allowing misinformation about vaccines and stuff like that, they already picked positions.** I, honestly the first thing I do is outlaw smear campaigns. Outright, just misleading voters, and preventing what's best for everybody. [emphasis added]

7:33 Ryan: But do you think that they're actively censoring like conservatives' voices? Facebook? [emphasis added]

7:36 Daniel: "No I mean a lot of them already violated policies either with harassment or stuff like that. The easiest one to point out is what's his name, Alex Jones. He was already banned on Facebook for harassment. He was banned for hate speech, he was banned because he kept calling the shooting fake and kept harassing family about it. And that's why he got a lawsuit, and **so you know what he crossed a line.** [emphasis added]

[Later on . . .]

8:30 **Ryan:** "So what you're saying is, to summarize, Facebook is, they do, Facebook does have like a bias, but it's not because of what you would think. The reason behind it isn't because of what you'd think. And they lean which way? Like toward?

8:42 **Daniel: I think they lean more toward, they lean the same way Google leans, which is more towards the left. Whatever keeps them safe. If the left . . . they're not wrong, the left will keep them safe because mass majority appeal on the Internet is left,** and that's how it's always gonna be until you know. Because they may be the little guys, but little guys on the Internet make a big voice. [emphasis added]

9:07 **Daniel:** That's the only thing I don't like is the whole cancel culture. Oh he's racist, let's get him fired from his job. Like why would you get him fired from his job?

9:12 **Ryan:** How does that help anybody?

9:13 **Daniel:** His personal views are his personal views.[5]

Paul is a coworker who I enjoyed talking to about politics. He was very insistent that he was politically neutral and independent, and he told me he analyzed the news and tried to look at things from an unbiased perspective.

Paul Deering,[5] audio only recording from 11/19/2019 Tuesday, around 11:30 a.m. outside the 2510 building.

---

5 A pseudonym.

0:00 **Paul:** . . . watch a lot of that stuff. Both sides do it and like it's so weird how one is just so facts and the other one is just so opinion.

0:06 **Ryan:** Yeah, like if you're truly unbiased about it, hey Liz. Yeah if you're truly unbiased, which way are you going?

0:13 **Paul:** I'm heading this way.

0:15 **Ryan:** If you're truly unbiased then like you can see the difference.

0:19 **Paul:** Yeah, it's so weird I mean.

0:22 **Ryan:** Just look at the facts.

0:24 **Paul:** And like I have views on both sides like both parties but it's like I look at 'em, some of 'em, I'm like man you guys, some of the stuff y'all pushing, like I don't know if this is working.

0:35 **Ryan:** But you think it's pretty clear, I mean with social media, like Twitter, Facebook, YouTube, they're all like biased towards the left?

0:40 **Paul: Oh yeah, that's what I tell people too, I'm like, I'm like, yo, movies, media, film, college, everything's to the left, like it's all pushed to the left**, I'm like, you know there's not really a, I'm like what's the right have? And they still are winning. Like I don't know. I don't know what else they have like really. Everything's for the most part being pushed on the other side so I'm like yeah I don't really know. And yeah they still like kick ass and win I don't know. But I think it's just more, cuz it tones to American traditional values and people on the right side, most Americans . . . [emphasis added]

1:15 **Ryan:** They still have values.

1:17 **Paul:** They're more to, and it works, it's what's worked for us, you know it's like, you look at the other countries, it's like our core values low key are what keep us to where we are. Yeah the more we try to be like these other countries, the more shit starts falling apart.[6]

11/20/2019 Paul Deering

0:32 (2019/11/20 00:45:18) **Ryan:** Do you think Facebook's gonna like try to make sure Trump doesn't win again? In the election?

0:39 **Paul:** Hmm, I mean if you can play a part in that, you know what I'm saying, but I don't know. I think . . . hmmm . . . that's tough.

0:52 **Ryan:** Based on what you've seen, like how they tell us to action certain things, do you feel like they are taking sides?

0:59 **Paul:** Oh yeah there's definitely a side-taking.

VIDEO 10 (2019/11/20   00:45:51)

0:02 **Ryan:** Towards the left?

0:03 **Paul:** Yeah, 100 percent. I mean it's media though. I mean it makes sense but you know they kind of control the media and run it, but I don't know that they, it's weird. I'm starting to see a weird. It's starting to lose its. It's almost like cry wolf. You know, like. So after so many times, like people aren't buying that shit no more.

0:23 **Ryan:** They can see through it.

0:25 **Paul:** They can see through the shit real quick.

VIDEO 12 Paul Deering 11/20/2019 Continued

0:01 **Paul:** [MORE DISCUSSION OF IMPEACHMENT]

0:15 **Ryan:** yeah but it's pretty clear yeah that like Facebook is like, so you think Facebook is like taking sides as far as I mean?

0:21 **Paul:** Like when I got this job I could definitely see the push to the left hard, um, just like with certain people being thrown on the hate list, and I was like oh really, huh. Interesting.

0:31 **Ryan:** Some people who are like reputable journalists who aren't racist.

0:33 **Paul:** Yeah who aren't racist but have just controversial subjects.

0:35 **Ryan:** Like Infowars.

0:38 **Paul:** Yeah when they got thrown out I was like, woah, like that's interesting.

0:43 **Ryan:** It was so ironic like, you're banning InfoWars, like that's what an Info War is, when you ban the other person.

0:50 **Paul:** Yeah facts. Alex Jones I get like . . . [7]

I've included a few more transcripts of conversations with coworkers below. I had a chat with Rafael Santana[**] about how the AI can filter things out and prevent hate speech before it happens. We talk about conspiracy theories and how Facebook screens out misinformation. The term used by Facebook for this is "proactive enqueuing." We talk about how that was like the movie *Minority Report*, where crimes are prevented before they are committed. In this conversation, we also talk about Facebook's stance on abortion.

1/25/2020 Rafael Santana

Video 10
3:35 2020/01/25 12:39:38'
Ryan: Rafael what do you think about that post about proactive enqueueing?

4:07 **Rafael:** I was reading that, and it's kind of weird.

2020/01/25 12:41:39
0:34 **Rafael:** But I'm thinking they can isolate individuals page or profile or whatever, and then use AI to filter out certain words that come up. Like isolated, filtered, directly, to get to it. Cuz if somebody doesn't report it then they're not gonna see it. So it's kind of like a self, like you said, minority report, essentially, because it's probably already there . . .

1:36 **Rafael:** It sounds like more they're waiting for people to report it, because it might not[violate]. . . . going right in and cleaning house.

** A pseudonym.

2:18 **Rafael:** I always thought we should have had a department that goes in and reports stuff . . . you don't catch every single violation. It gets reported out there. But that's only 50% of what's really violating.

2:47 **Ryan:** If we see something that's misleading, like we definitely know that that's false, what would we do? [Later on . . .]

3:14 **Ryan:** Should they implement fact checking?

3:23 **Rafael:** It should, but what would be the turnaround time?

3:35 **Ryan:** But it should be a priority, because you wanna protect, you wanna make sure people aren't being—

3:40 **Rafael:** You wanna protect the integrity of what's right and wrong. Or what's correct or not.

3:45 **Ryan:** So people aren't exposed to false information.

3:56 **Rafael:** Cuz you have a lot of conspiracy sites that are up that are still flourishing. Like the moon landing.

4:15 **Ryan:** Cuz they shouldn't be allowed to spread stuff because it's not like a free platform, it's not like the Internet, it's a private—

4:35 **Ryan:** But who decides what's false or not? So the moderator basically decides what the user gets to see, right?

4:57 **Rafael:** Yeah. That conspiracy stuff, **I'm kind of open to leaving it up, like people make their own decisions.** [emphasis added]

0:09 **Ryan:** But what about vaccines? What about abortion? What if there are facts that are misleading about abortion? What would you do?

[Long Pause]

0:24 **Rafael: Touchy. It's like do you go medically or do you go factually? It's the heart and head kind of thing.** [emphasis added]

0:41 **Ryan:** But you want to make sure there's not misinformation. You want to help young women who are making choices about their body, right?

0:52 **Rafael: Having people like us trying to decipher stuff like that is a lot of dissecting. Kind of what we do now anyways.** [emphasis added]

1:09 **Ryan: In a way would you say we're like human filters?** [emphasis added]

1:10 **Rafael: A little bit. A lot of influence.** [emphasis added]

1:25 **Ryan:** That's why they depend on their AI so much.

1:50 **Ryan:** Facebook has their own terms, their own glossary . . .

2:20 **Rafael:** It's kind of like lawyer-speak.

3:02 **Ryan: Technically per the jargon an abortion isn't a violent death.** [emphasis added]

3:07 **Rafael:** Yeah.

2020/01/25 12:49:08

3:12 **Rafael:** So you kind of already know. It doesn't define where they stand, but they needed to stand somewhere. To be honest.

3:18 **Ryan: Do you think that's an issue though, that they're defining it in a different way?** [emphasis added]

3:23 **Rafael: Yeah because you can't apply cruel [cruel and insensitive policy]. You can mock a fetus.** [emphasis added]

3:29 **Ryan:** What? Cruel & Insensitive?

3:31 **Rafael: Yeah, you can mock it. You can make fun of it. But it doesn't seem right.** [emphasis added]

3:39 Ryan: Because they're like it's not violent, but it is violent, but because it doesn't fit their criteria for—

3:55 Rafael: They don't consider them [fetuses] a person. Living person. So you go back to the do you have a heartbeat or not . . . [emphasis added]

4:09 Ryan: So they're deciding in a way. Facebook's deciding what?

4:15 Rafael: I don't know about deciding, but they have to stand somewhere. It might not be the popular one, but—

4:28 Ryan: But they're making a decision about what that means. What abortion means, right?

4:34 Felipe: Maybe.[8]

The following is a conversation with my team leader, Mario Rodriguez,[††] who was my superior while I was on the North America side. We talk about Facebook's policy and whether it has bias.

12/20/2019 Conversation with Team Leader Mario Rodriguez

1:18 Ryan: I guess they started fact-checking earlier this year in like the UK. But um I mean if they're fact-checking—

1:25 Mario: Facebook?

1:28 Ryan: Yeah they hired out contractors to do it.

1:30 Mario: Sounds like a hard job.

1:30 Ryan: But would they be biased though towards a certain—

1:34 Mario: Nah, cuz I'm pretty sure they follow policy.

1:38 Ryan: But who designs the policy?

---

††A pseudonym.

1:39 **Mario:** Facebook, but yeah I can see your point there. That's actually a good point. But I think it would be. I don't think so. I'm pretty sure there's people intelligent enough to create policies that wouldn't sound biased.

1:55 **Ryan:** Yeah.

1:58 **Mario:** I would like to see those policies though. I don't really keep up with politics, do you?

2:01 **Ryan:** Yeah, quite a bit. So you see different discrepancies where it's like, okay they're protecting this Greta Thunberg person who's a minor, but then earlier this year there was this Trump supporter outside of the Lincoln Memorial who got like targeted and made fun of and he was like 17, so, and there was like no protections for him.

2:22 **Mario:** So there's bias.

2:24 **Ryan:** Yeah, I mean from what you've seen, do you think there's bias?

2:27 **Mario:** Definitely dude.

2:30 **Ryan:** I mean we can be honest about it right?

2:31 **Mario:** Yeah.[9]

Another member of leadership I talked to was Nick Crosnan.[‡‡] I hadn't interacted with him very much, but he had been a team leader and then transitioned to Workforce Management (WFM), basically tracking all of the employees' statistics and reporting it to Facebook. I went down to the fourth floor and had the following conversation:

12/20/2019 Nick Crosnan, former Team Leader

3:15 **Ryan:** [mentions elections . . . how Facebook brought everything to the U.S. . . . someone in India is not gonna understand Hillary and Trump and all the nuances]

---

‡‡A pseudonym.

3:32 Nick: Yeah even us, even us we had to do, cuz they started doing, remember they sent people over to Texas, to Austin, to prep 'em for the election. So they got like insight on that, and they were supposed to be POCs, specifically for that stuff. So yeah cuz even us we had to get, you know, kind of like, refreshers, and you know, start doing research.

3:52 Ryan: But some of the stuff makes me question, they made white trash an ignore, calling cops pigs is allowed now.

4:15 Ryan: you wonder if there's some bias going on. **How much skin does Facebook have in the game for the election? Are they gunning for one candidate or another?** [emphasis added]

4:26 Nick: We mostly can tell when they make those, um, like on the chats. Cuz I'm still on like a lot of the workplace chats. When they make like exceptions, they're like, hey we have no clue what's going on, but we're gonna, you know, action stuff this way. **So that kind of lets you know where they are, it's when they have to make those quick decisions.** Cuz they're like hey we don't know. And then sometimes they'll change it. Like they'll allow something, and then a week later it's now a delete. **But it's that initial response that they're like, this is how we're initially gonna handle it. Kind of lets you know where they're lying.** [emphasis added]

Here is yet another conversation with someone in a leadership position. I ask him if he thinks Facebook is trying to push a certain candidate for the 2020 presidential election:

1/18/2020 Javier Romero,[§§] Team Leader

2:10 Ryan: Do you think the election is gonna be like, what do you think Facebook's doing to like make sure crazy people aren't advocating for violence, I mean we have a lot of policies now that keep that from happening, we have the violence and incitement policy.

2:29 Javier: I think maybe we're doing like a little bit to help, I guess, protecting people, with like people advocating or aligning with, even when Solemani was irrelevant, he still is, but a lot of that for PSR. And just advocating for death on Trump and shit like that.

[§§] A pseudonym.

2:53 **Ryan:** They could do more right? Facebook could do more to—

2:55 **Javier:** Yeah, yeah, we could do a lot more. But I mean guess that's why we have the LERT [send to law enforcement] or whatever, but that's only for like more CEI [Child Exploitative Imagery] and shit like that so, which I think it should really extend, past that into dangerous orgs, just cuz of everything that's going on. But I mean.

3:16 **Ryan:** Do you think Facebook's trying to make sure that a certain person wins in November? Or, I don't know, do you think they're gonna try to root for.

3:25 **Javier: I'm convinced it's biased, as far as social media goes. Being such big platforms, and being able to control what's going on and off the platform.** I think really does something to effect of it. Especially during campaigns and stuff like that. Definitely. Try to lower influence because of they get pros [advantages] out of it, or who knows. But I think yeah, definitely social media dictates that for sure. Mainly Twitter. Trump I think is a huge one.[10] [emphasis added]

I had many conversations with Shawn Browder, the policy manager, and on this occasion another person was sitting at the table when I came to talk to him. Her name is Heidi Henrickson,[¶¶] and she was involved in planning a lot the activities for the employees, such as a wiffle ball tournament or other events to improve morale.

02/04/2020 Conversation with Policy Manager Shawn Browder and Heidi Henrickson Video 24
Timestamp   2020/02/04 13:56:05

1:14 **Ryan:** What's up Shawn?

[We talk about the project ending.]

1:24 **Ryan:** People are saying, the last day, I'm gonna ignore everything [leave content on the platform even if it violates].

1:25 **Shawn:** [laughs hysterically]

¶¶ A pseudonym.

1:38 Shawn: Just leave everything on the platform.

1:39 Ryan: it's either delete everything or ignore everything.

[Discussion about guidance from Facebook about the Coronavirus.
People could think bleach is a cure for Coronavirus.]

3:25 [Only other thing was the Iowa Caucus—discussion regarding Russian Interference.]

Timestamp 2020/02/04 13:58:45

4:00 Shawn: She was reading on *The Onion* that they're crowdfunding 500 million for a pen that would count votes.

4:25 Ryan: Was Facebook giving any guidance about the Iowa Caucus?

4:35 Shawn: Just standard stuff, hey, be on the lookout.

4:53 Ryan: I don't think the DNC is gonna want Bernie.

4:59 Shawn: I know they didn't want him.

Video 25
Timestamp   2020/02/04 13:59:53

0:01 Shawn: I know they didn't want him but here's the thing, if you are sandbagging the candidate that your biggest voter bloc wants, you're not gonna have a voter bloc that supports you candidate.

[Heidi claps in support]

0:23 Shawn: Like stop sandbagging millennials, millennials are your biggest voting bloc. Cater to us please. [emphasis added]

0:32 Heidi: It's about god damn time. [emphasis added]

[Heidi talks about how the DNC wants Biden.
If you get Biden, then you've handed the presidency to Trump.
Hillary is establishment legacy politics.

There's too much sexism for people to elect Elizabeth Warren. (1:12)]

1:20 **Heidi:** (Warren won't win the rust belt because they're too sexist.

1:22 **Ryan:** Yeah Facebook should have their own little pro-Bernie squad.

2:08 **Ryan:** At least Bernie sticks to his guns.

2:25 **Ryan:** Some of [Bernie's] policies are more left than what Obama ran on, maybe?

2:31 **Shawn:** Yes, however what's he's proposing in line with literally the rest of the western world. [Europe, Canada]

2:48 **Heidi: His [Bernie's] proposals are centrist for the rest of the world . . . of countries that are already past this. What he's bringing up is not crazy. This is just where you develop next.** [emphasis added]

3:12 **Ryan: Wouldn't Facebook want Bernie to win then?** [emphasis added]

3:15 **Elizabeth: Oh no, definitely not. He's not a monopoly person. Google, Uber, none of them like him.**[11] [emphasis added]

END OF INTERACTION

Jessica Alvarez*** is the team manager for the North America side. Her superior is a service delivery manager, who then reports to Jim Denton,††† the site lead. Jessica supervises about three or four team leaders.

Previously, Jessica was the team leader for the PDO team, which deals with hate speech. She interacted on a weekly basis with the client during conference calls while working in PDO. She says in the video that there are about fifteen members of the policy team at Facebook who decide everything, and she thinks there's bias amongst them.

*** A pseudonym.
††† A pseudonym.

12/11/2019   Interaction with Alexxas Alvarado, Team Manager

6:01 **Ryan:** Yeah, but based on, I mean, do you interface with the client? With Facebook a lot, or not?

6:08 **Jessica:** No, not since moving to NA.

6:14 **Ryan:** Where were you before?

6:15 **Jessica:** PDO, and those were like weekly calls, that I would be on but since I've been here I'm not involved in those calls.

6:24 **Ryan:** PDO is like hate speech, kind of?

6:25 **Jessica:** Uh huh, hate speech, so, well it was hate speech, now it's like broken into different pieces since I came up here, so they have descriptive labeling, I don't know what it entails. But for the most part it was like hate speech and you'd see a lot of political stuff, um, for like all over. So you'd be responsible for like hitting the people, I don't know, someone made like a powerpoint, and there was like all these little people, and I'm like, I don't know all you guys can remember it.

6:58 **Ryan:** Did you see anything when you were in PDO as far as like, hey they're kind of like leaning towards one side, or favoring like leftist content or censoring kind of . . .

7:10 **Jessica:** No because it would be the same, it's the same content that you guys see, so the content in PDO has already gone through CO, and PDO is looking at false positives, yeah I think they're looking at false positives. False positives, I wanna say false positives. **To see like how you guys miss stuff, and how can they teach an AI not to.** [emphasis added]

7:45 **Ryan:** Yeah I had a friend in PDO, I recruited him, and I referred him, but yeah with the 2020, like we get so much politics, and **we had like the whistleblower [Ciaramella], and how they're like protecting him, and not letting his name be published.** So I know you're not too political, but from what you've seen, do you think like, do you think Facebook's like leaning or biased towards the right or the left? [emphasis here]

8:12 Jessica: I don't know, like so I don't follow a lot of that stuff. So I know there was a whistleblower, I don't know exactly what he had given up and stuff like that, **but I was in on that call, like where they had said they removed everything from the platform, regarding him. You couldn't post his name or anything. Cuz to me it's like, why would you remove it? Like his name's going to be published like everywhere else, why not just leave it on the platform.** I did like remove [unintelligible] the negative stuff about him. But if it's like really just informative, what's the purpose of removing it? [emphasis added]

8:52 Ryan: And you wonder if the roles were reversed, like if it were a pro-Trump whistleblower, or if like, would they go to the same lengths to protect him right?

9:00 Jessica: I remember in class a few weeks ago we were talking about like who decides all of these things when it comes to what's allowable on the platform. We went into this from like automated cars. But who decides like, what's allowable and stuff and like, it comes into morals and stuff like that, and really when you have, I wanna say it's like 15 people who do the policy and then like obviously it trickles down. [emphasis added]

9:26 Ryan: At Facebook?

9:27 Jessica: Yeah, and then it trickles down to like the market teams, and stuff like that, but when you have like a set of people deciding this, you can see where there would be biases and stuff like that, even if it is unconscious bias. And when you tend to work with people for a long period of time you guys start to think alike and stuff like that, without even knowing. To say there wouldn't be any bias at all. [emphasis added]

9:58 Ryan: It's kind of hard to avoid that yeah.[12]

As you know, Israel Amparan appeared in the Project Veritas video, and he was quite open about his hate for Trump. Here is another conversation I had with him along with the person sitting next to him, named Amanda. I do agree with Israel that it shouldn't be Facebook's job to police what people say, and Israel admits it's a public platform. However, I disagree with most of his other points.

12/11/2019 Interaction with Israel Amparan and Amanda Iberio‡‡‡

Video 7

7:18 **Israel:** Facebook's getting too much fucking criticism for that shit, like it's not Zuckerberg's fucking job to police what people say in a public fucking platform. You know, and it's like, if you wanna get your news from an irreputable source, then that's your problem, no matter which way you fucking slice it.

7:34 **Amanda: But the problem is that people think that like Facebook's credible, you know what I mean?** [emphasis added]

7:40 **Israel:** Yeah but that's on them, not on Facebook

7:44 **Ryan:** But it's like the whistleblower, so we're deleting the Eric Ciaramella post right?

7:48 **Israel:** Oh, that makes me so mad every time I delete that post.

7:51 **Ryan:** So that's good, but let's be honest, if the roles were reversed, if it were a pro-Trump whistleblower, do you think that Facebook would be protecting them as much?

7:56 **Israel:** I think so.

7:58 **Amanda:** I don't know.

8:03 **Israel:** I like to think that it's just based off their status as a vulnerable person, not so much as whom they're blowing whistles against, you know, **but then you know, maybe a pro-Trump whistleblower wouldn't need nearly the amount of protection this guy needs.** You know, but Trump supporters are fucking crazy ass assholes, that every other fucking word out of their mouth is you know "come take it," "seal the border," like listen, I hate government as much as the next fucking person, but you're not gonna catch me riding over the fucking, it's like impeachment, it's like a [unintelligible] leg, it doesn't just happen. Trump called it a fucking coup. And it's like that should scare you more than anything. [emphasis added]

‡‡‡ A pseudonym.

8:40 Amanda: You know what makes me sooo angry is that a lot of people think that Antifa like represents all the like liberals and all of the new ideas, LGBT, they do NOT represent that at all. I don't agree with anything. [emphasis added]

8:44 Ryan: They're like fringe.

8:59 Amanda: They're like super far-off, it's like almost to the point where they're like hypocrite.

Video 8

1:50 Ryan: At the end of the day (I'm gonna head out in a second), but at the end of the day, even if Trump was being censored, or like shadowbanned, or if conservatives were being shadowbanned by Facebook, wouldn't that be like a good thing? Cuz like, some of these people are crazy.

2:06 Israel: Yeah, no honestly [nods head], like every time I, like half the time when I delete people for like Trump shit, I'm like, you should be on a watch list dude. You know, Trump supporters like to throw around like Trump Derangement Syndrome, as like, you know liberals being crazy. I like to think of it as, actually you're the one that has fucking Trump Derangement Syndrome, because you're losing your goddamn mind as soon as, cuz they're just like "Oh Trump 2020" [unintelligible], and it's like, that sounds a little more deranged to me when you end every argument in that. [emphasis added]

2:39 Ryan: Yeah both sides are crazy when it comes to politics.

2:44 Israel: The stakes are just high, right now. You know. I don't think politics has been as impactful as they are now. **And Trump who's constantly instigating and stoking the fires, and using this heavy, loaded rhetoric.** And that's what he does is he uses rhetoric that has a lot of like kind of subliminal fucking messages and a lot of undertones, to sway what you think, and all his stuff, like you know when they're attacking me they're attacking you, basically really trying to put the whole, he's really trying to put himself in the center of everything . . . [emphasis added]

3:17 Amanda: He's trying to make it look like he IS the people.

3:22 Israel: The president isn't America, it's a singular branch or office, like oh you have to respect our president. No I don't. It's actually more American to disrespect him than to respect him.

END OF INTERACTION[13]

Acknowledgments

From Ryan:

I would like to thank Constance Hathaway for volunteering her time to help me review each chapter to ensure they were polished.

I'd also like to thank Project Veritas and their team for their logistical support while I was filming with a hidden camera.

A huge thank-you to my literary agent, Johanna Maaghul, who moved heaven and Earth to bring to fruition this book deal.

I couldn't have written this book without the tremendous vision and literary abilities of my coauthor, Kent Heckenlively. Thank you, Kent, for helping the story take shape and bringing it to life.

All of this couldn't have taken place without the support of my wife Livy, who knew the risk I was taking and was always supportive of my decision to become a whistleblower.

From Kent:

I'd first like to thank my wonderful partner in life, Linda, and our two children Jacqueline and Ben, for their constant love and support. I'd like to thank my mother, Josephine, and my father, Jack, for teaching me to tell the truth regardless of the consequences and showing how to love through even difficult times. I'd like to thank the best brother in the world, Jay, and his wonderful wife, Andrea, and their three children, Anna, John, and Laura, for always being on my side.

I'd like to thank some of the wonderful teachers in my life, my seventh-grade science teacher, Paul Rago, my eighth-grade English teacher, Elizabeth White, my high school science teacher, Ed Balsdon, my religion teacher Brother Richard Orona, and in college, English professors Clinton

Bond, Robert Haas, Carol Lashoff, and in the political science department, David Alvarez, who nominated me to be the school's Rhodes Scholar candidate. I'd also like to thank my college rowing coach, Giancarlo Trevisan, the mad Italian, who showed me what it means to have crazy passion for an often-overlooked sport. In law school, I'd like to thank Bernie Segal, the criminal defense attorney who taught me to always have hope that justice will eventually prevail. I'd like to thank my writing teachers, James Frey, who looked at me one time and said, "Yeah, I think you'll be a writer," as well as Donna Levin, and James Dalessandro, who always said to find the story first, then write the hell out of it.

My life wouldn't be complete without my great friends, John Wible, John Henry, Pete Klenow, Chris Sweeney, Suzanne Golibart, Gina Cioffi, Eric Holm, Susanne Brown, Rick Friedling, Max Swafford, Sherilyn Todd, Rick and Robin Kreutzer, Christie and Joaquim Pereira, Tricia Mangiapane, and all of you who have made my passage through life such a party.

I work with the best group of science teachers at Gale Ranch Middle School, Danielle Pisa, Neelam Bhokani, Amelia Larson, Matt Lundberg, Katie Strube, Derek Augarten, and Arash Pakhdal. Thanks for always challenging my thinking and making me ask what is best for our students.

In the activist community, I'd like to thank J.B. Handley, Tim Bolen, Mary Holland, Lou Conte, Del Bigtree, Brian Hooker, Barry Segal, Elizabeth Horn, Brian Burrowes, Polly Tommey, Dr. Andrew Wakefield, and Robert F. Kennedy, Jr., for their friendship in the continuing fight against the Goliath of corrupted science.

Lastly, I'd like to thank my agent, Johanna Maaghul, my wonderful editors, Anna Wostenberg, and at Skyhorse the fabulous Caroline Russomanno, and for the faith shown in me over the years by publisher Tony Lyons.

Endnotes

CHAPTER 1

1 Casey Newton, "The Trauma Floor: The Secret lives of Facebook Moderators in America," "The Verge," February 25, 2019, //www.theverge.com/2019/2/25/18229714/cognizant -facebook-content-moderator-interviews-trauma-working-conditions-arizona.

2 "Gospel Topics—Profanity," The Church of Jesus Christ of Latter-Day Saints (Accessed March 9, 2021), www.churchofjesuschrist.org/study/manual/gospel-topics /profanity?lang=eng.

CHAPTER 3

1 Nick Hopkins, "Revealed: Facebook Internal Rulebook on Sex, Terrorism and Violence," *The Guardian*, May 21, 2017, www.theguardian.com/news/2017/may/21/revealed -facebook-internal-rulebook-sex-terrorism-violence.

2 Ibid.

3 Ibid.

4 Ibid.

5 Ibid.

6 Transcript of screen filmed on Project Veritas's behalf, November 6, 2019.

7 Transcript of video taken on Project Veritas's behalf, December 26, 2019.

8 Casey Newton, "The Trauma Floor: The Secret lives of Facebook Moderators in America," "The Verge", February 25, 2019.

CHAPTER 4

1 Casey Newton, "The Trauma Floor—The Secret Lives of Facebook Moderators in America," "The Verge," February 25, 2019, www.theverge.com/2019/2/25/18229714 /cognizant-facebook-content-moderator-interviews-trauma-working-conditions-arizona.

2 Ryan Hartwig, "Letter to Various Congressional Members, United States Senators, and Journalists," May 14, 2019.

3 Ibid.

4 Ibid.

5 Ibid.

[6] "Telephone Conversation of President Trump with President Zelensky of Ukraine," July 25, 2019 (Declassified September 24, 2019), www.games-cdn.washingtonpost .com/notes/prod/default/documents/d2311f4f-a767-4ddc-868b-8bc9af8226c5 /note/339b784b-719c-464f-9eda-85daede53092.pdf#page=1.

[7] Transcript of video taken on Project Veritas's behalf, November 8, 2019.

[8] Ibid.

[9] Transcript of video taken on Project Veritas's behalf, November 13, 2019.

[10] Transcript of video taken on Project Veritas's behalf, November 19, 2019.

[11] J.D. Rucker, "Can We Talk About Eric Ciaramella?" Real Clear Politics, February 16, 2020, www.realclearpolitics.com/2020/02/16/can_we_talk_about_eric_ciaramella_501613 .html.

[12] Ibid.

[13] Alexander Vindman, "Opinion: Alexander Vindman: Coming Forward Ended My Career. I Still Believe Doing What's Right Matters," Washington Post, August 1, 2020, www.washingtonpost.com/opinions/2020/08/01/alexander-vindman-retiring-oped/.

CHAPTER 5

[1] "Mark Twain Quotes," GoodReads (Accessed March 23, 2021), www.goodreads.com /quotes/78468-when-i-was-a-boy-of-14-my-father-was.

[2] Kyann-Sian Williams, "Read Climate Change Activist Greta Thunberg's Incredible Speech to the UN in Full," NME, September 24, 2019, www.nme.com/features/greta -thunberg-full-speech-to-the-un-2019-climate-change-summit-2550824.

[3] Ibid.

[4] Karen Tumulty, "Commentary: Greta Thunberg vs. Donald Trump," Chicago Tribune, December 12, 2019, www.chicagotribune.com/opinion/commentary/ct-opinion-greta -thunberg-trump-time-20191212-kludyj3mnfa25pruin5h52tcde-story.html.

[5] Charlotte Alter, Suyin Haynes, and Justin Worland, "TIME Person of the Year—Greta Thunberg—The Power of Youth," TIME, December 23, 2019, www.time.com/person -of-the-year-2019-greta-thunberg/.

[6] Ibid.

[7] Ibid.

[8] Robert Rapier, "China Emits More Carbon Dioxide Than The U.S. and EU Combined," Forbes, July 1, 2018, https://www.forbes.com/sites/rrapier/2018/07/01/china-emits -more-carbon-dioxide-than-the-u-s-and-eu-combined/?sh=3ab8941a628c.

[9] Ibid.

[10] Charlotte Alter, Suyin Haynes, and Justin Worland, "TIME Person of the Year."

[11] Transcript of video taken on Project Veritas's behalf, December 13, 2019.

[12] Transcript of screen filmed on Project Veritas's behalf, October 3, 2019.

[13] Transcript of screen filmed on Project Veritas's behalf, December 24, 2019.

[14] Transcript of screen filmed on Project Veritas's behalf, December 13, 2019.

[15] Transcript of screen filmed on Project Veritas's behalf, September 24, 2019.

CHAPTER 6

[1] Craig Silverman, Ryan Mac, and Pranav Dixit, "'I Have Blood on My Hands': A Whistleblower Says Facebook Ignored Global Political Manipulation," Buzz Feed News,

September 14, 2020, www.buzzfeednews.com/article/craigsilverman/facebook-ignore-political-manipulation-whistleblower-memo.

2 Ibid.

3 Ibid.

4 Ibid.

5 Ibid.

6 Transcript of screen filmed on Project Veritas's behalf, February 5, 2020.

7 Miranda Devine, "Project Veritas uncovers 'ballot harvesting fraud' in Minnesota: Devine," *New York Post*, September 27, 2020, https://nypost.com/2020/09/27/project-veritas-uncovers-ballot-harvesting-fraud-in-minnesota/.

8 "Omar Connected Harvester SEEN exchanging $200 for General Election Ballot. 'We don't care illegal'," Project Veritas, YouTube (Accessed April 17, 2021, https://youtu.be/MV7oDl8yDZk.

9 Haley Victory Smith, "Texas woman arrested on voter fraud charges," *Washington Examiner*, January 13, 2021, https://www.washingtonexaminer.com/news/texas-woman-arrested-voter-fraud-charges.

10 Transcript of screen filmed on Project Veritas's behalf, November 6, 2019.

11 Transcript of screen filmed on Project Veritas's behalf, December 12, 2019.

12 Transcript of screen filmed on Project Veritas's behalf, November 6, 2019.

13 David Smiley and Alex Daughtery, "DeSantis says Florida shouldn't 'monkey this up' by electing Andrew Gillum," *Miami Herald*, August 29, 2018, https://www.miamiherald.com/news/politics-government/state-politics/article217507400.html.

14 Transcript of screen filmed on Project Veritas's behalf, November 6, 2019.

15 Ibid.

16 Ibid.

17 Ibid.

18 Ibid.

19 Transcript of video taken on Project Veritas's behalf, October 10, 2019.

20 Margot Cleveland, "Lawsuit Exposes How the Media and Deep State Hatched The Russiagate Hoax," *The Federalist*, May 31, 2019, https://thefederalist.com/2019/05/31/lawsuit-exposes-media-deep-state-hatched-russiagate-hoax/.

21 Transcript of screen filmed on Project Veritas's behalf, June 6, 2019.

22 Felix Allen, "Catalan leader Carles Puigdemont flees to Brussels amid fears Spain could jail him for 30 years for 'rebellion' over independence row," *The Sun*, October 30, 2017, https://www.thesun.co.uk/news/4801211/carles-puigdemont-independence-catalan-leader-flees-spain-to-brussels/.

23 Transcript of Canadian election powerpoint filmed on Project Veritas's behalf, September 24, 2019.

24 Ibid.

25 Ibid.

26 Ibid.

27 Transcript of screen filmed on Project Veritas's behalf, June 6, 2019.

28 "What is Venezuela's Bolivarian Revolution?" TRT World, January 24, 2019, https://www.trtworld.com/americas/what-is-venezuela-s-bolivarian-revolution-23587.

29 "Video Shows Notorious 'Colectivo' Leader Greeted by Venezuela Officials," *Insight Crime*, February 28, 2018, https://insightcrime.org/news/brief/suspected-murderer-colectivo-leader-greeted-venezuela-officials/.

30 Transcript of screen filmed on Project Veritas's behalf, October 12, 2019.
31 Transcript of screen filmed on Project Veritas's behalf, June 6, 2019.

CHAPTER 7

1 "Number of daily active Facebook users worldwide as of 4th quarter 2020," *Statista*, January 2021, https://www.statista.com/statistics/346167/facebook-global-dau/.
2 "Facebook for government, politics and advocacy," Facebook, accessed April 17, 2017, https://www.facebook.com/gpa.
3 Craig Silverman, Ryan Mac, and Pranav Dixit, "'I Have Blood on My Hands': A Whistleblower Says Facebook Ignored Global Political Manipulation," *Buzz Feed News*, September 14, 2020, www.buzzfeednews.com/article/craigsilverman/facebook-ignore-political-manipulation-whistleblower-memo.
4 Ibid.
5 Ibid.
6 "Tech CEOs Senate Testimony Transcript October 28," (Accessed April 17, 2021), https://www.rev.com/blog/transcripts/tech-ceos-senate-testimony-transcript-october-28.
7 Heather Kelly, "Hate groups on Facebook: Why some get to stay," CNN, August 17, 2017, https://money.cnn.com/2017/08/17/technology/culture/facebook-hate-groups/index.html.
8 Kyle Smith, "Hate, Inc.: The SPLC Is a Hyper-Partisan Scam," *National Review*, March 1, 2018, https://www.nationalreview.com/2018/03/southern-poverty-law-center-bias-hate-group-labels-scam/.
9 Jessica Prol Smith, "The Southern Poverty Law Center is a hate-based scam that nearly caused me to be murdered," *USA Today*, August 17, 2019, https://www.usatoday.com/story/opinion/2019/08/17/southern-poverty-law-center-hate-groups-scam-column/2022301001/.
10 Craig Silverman, Ryan Mac, and Pranav Dixit, "'I Have Blood on My Hands'." September 14, 2020.
11 Nick, "Facebook Censors Polish Nationalists," Research Teacher, October 30, 2016, https://researchteacher.com/facebook-deleting-polish-nationalist-pages/.
12 Eric Lieberman, "Hungarian PM's Chief Of Staff Says Facebook Censored His Anti-Immigrant Video," *Daily Caller*, March 8, 2018, https://dailycaller.com/2018/03/08/hungary-janos-lazar-facebook-censored-video/.
13 Emma Graham-Harrison and Sam Jones, "Facebook takes down far-right groups days before Spanish election," *The Guardian*, April 25, 2019, https://www.theguardian.com/world/2019/apr/25/facebook-takes-down-far-right-groups-days-before-spanish-election.
14 Julio Ariza, "Censura en las RRSS: Ryan Hartwig, moderador de contenido de Facebook," YouTube video, July 2, 2020, https://youtu.be/Jhi-khZB3s4.
15 Transcript of screen filmed on Project Veritas's behalf, June 6, 2019.
16 Estefania Arnaiz Mata, Facebook message to author, July 3, 2020.
17 Transcript of screen filmed on Project Veritas's behalf, June 7, 2019.
18 Ibid.
19 Edward White, "The Systematic Killing Of Christians Continues," ACLJ, April 23, 2015, https://aclj.org/persecuted-church/systematic-killing-of-christians-continues?view=designVariantB.

20 Josh Hanrahan and Stephen Gibbs, "Evil terrorist's twisted manifesto . . .," *Daily Mail*, March 16, 2019, https://www.dailymail.co.uk/news/article-6812183/Christchurch -terror-attack-Shooter-published-manifesto-online-hours-earlier.html.

21 "List of Islamic Terror Attacks on Christians," TROP (Accessed April 17, 2021), https:// www.thereligionofpeace.com/attacks/christian-attacks.aspx.

22 Tony Romm and Elizabeth Dwoskin, "Facebook says it will now block white nationalist, white separatist posts," Stuff, March 28, 2019, https://www.stuff.co.nz /business/111604263/facebook-says-it-will-now-block-white-nationalist-white -separatist-posts.

23 Ebony Bowden, "Facebook to rank anti-black hate speech over anti-white comments," *New York Post*, December 4, 2020, https://nypost.com/2020/12/04/facebooks-hate-speech -algorithm-to-prioritize-anti-black-comments/.

24 Transcript of screen filmed on Project Veritas's behalf, October 12, 2019.

25 Ibid.

26 "Dying of Whiteness: How the Politics of Racial Resentment Is Killing America's Heartland," Amazon (Accessed April 17, 2021), https://www.amazon.com/Dying -Whiteness-Politics-Resentment-Heartland/dp/1541644980#ace-9766277718.

27 Notes taken based on documents filmed on Project Veritas's behalf, September 12, 2019.

28 Transcript of screen filmed on Project Veritas's behalf, June 6, 2019.

CHAPTER 8

1   Transcript of screen filmed on Project Veritas's behalf, October 17, 2019.

2   Transcript of screen filmed on Project Veritas's behalf, June 6, 2019.

3   Ibid.

4   Transcript of screen filmed on Project Veritas's behalf, June 7, 2019.

5   Ibid.

6   Transcript of screen filmed on Project Veritas's behalf, June 6, 2019.

7   Oliver Darcy, "Media outlets take Trump out of context to suggest he called undocumented immigrants 'animals'," CNN Business, May 17, 2018, https://money .cnn.com/2018/05/17/media/media-trump-animals-immigrants/index.html.

8   "Texas Criminal Illegal Alien Data," Texas Department of Public Safety (Accessed April 17, 2021, https://www.dps.texas.gov/section/crime-records-service/texas-criminal-illegal -alien-data.

9   Hans A. von Spakovsky, "Crimes by Illegal Immigrants Widespread Across U.S.— Sanctuaries Shouldn't Shield Them," The Heritage Foundation, September 3, 2019, https://www.heritage.org/crime-and-justice/commentary/crimes-illegal-immigrants -widespread-across-us-sanctuaries-shouldnt.

10 "John Stuart Mill: the Harm Principle," *Film and Philosophy* (Accessed April 17, 2021), https://filmandphilosophy.com/2013/03/08/john-stuart-mill-the-harm-principle/.

11 Transcript of screen filmed on Project Veritas's behalf, June 6, 2019.

12 Ibid.

13 Ibid.

14 Ibid.

15 Ibid.

16 Ibid.

17 Dr. Randall Price, "Examining the Significance of Jerusalem as Israel's Capital," *Liberty Journal*, February 21, 2018, https://www.liberty.edu/journal/article/examining-the-significance-of-jerusalem-as-israels-capital/.

18 Transcript of screen filmed on Project Veritas's behalf, September 12, 2019.

19 Ibid.

20 Transcript of screen filmed on Project Veritas's behalf, January 22, 2020.

21 Transcript of screen filmed on Project Veritas's behalf, September 17, 2019.

22 Transcript of 2018 Senate Hearing, WikiSource.Org (Accessed April 21, 2021), https://en.wikisource.org/wiki/Zuckerberg_Senate_Transcript_2018.

23 Transcript of screen filmed on Project Veritas's behalf, October 12, 2019.

24 Transcript of screen filmed on Project Veritas's behalf, February 26, 2020.

25 Transcript of screen filmed on Project Veritas's behalf, October 12, 2019.

26 Transcript of screen filmed on Project Veritas's behalf, October 3, 2019.

27 Transcript of screen filmed on Project Veritas's behalf, June 6, 2019.

28 Ibid.

29 Greg Norman, "Man accused of tossing drink at pro-Trump teen inside Whataburger arrested," Fox News, July 6, 2018, https://www.foxnews.com/us/man-accused-of-tossing-drink-at-pro-trump-teen-inside-whataburger-arrested.

30 Transcript of screen filmed on Project Veritas's behalf, February 26, 2020.

31 Ibid.

32 Renee Diresta, "Facebook's Anti-Conservative Bias Audit Is Here," *Slate*, August 21, 2019, https://slate.com/technology/2019/08/facebook-anti-conservative-bias-audit-results-kyl.html.

CHAPTER 9

1 Transcript of screen filmed on Project Veritas's behalf, February 26, 2020.

2 Ibid.

3 Ahmed Waqas et al., "Mapping online hate: A scientometric analysis on research trends and hotspots in research on online hate," *PloS ONE* (2019), 14(9): e0222194, https://doi.org/10.1371/journal.pone.0222194.

4 Transcript of screen filmed on Project Veritas's behalf, June 6, 2019.

5 Katja Rost, Lea Stahel, Bruno S. Frey, "Digital Social Norm Enforcement: Online Firestorms in Social Media," *PloS ONE* (2016), 11(6): e0155923, https://doi.org/10.1371/journal.pone.0155923.

6 Ibid.

7 Louise Hall, "'When our actions do not, our fears make us traitors': Ted Cruz trolled with Shakespeare quotes during literature dispute," *Independent*, February 11, 2021, https://www.independent.co.uk/news/world/americas/us-politics/ted-cruz-shakespeare-impeachment-twitter-andrea-mitchell-b1800852.html.

8 Andrea Mitchell (@mitchellreports), Twitter tweet, February 10, 2021, https://twitter.com/mitchellreports/status/1359650012125421572?s=20.

9 Charles C.W. Cooke, "No, It's Shakespeare," *National Review*, February 10, 2021, https://www.nationalreview.com/corner/no-its-shakespeare/.

10 Transcript of screen filmed on Project Veritas's behalf, February 26, 2020.

CHAPTER 10

1   Valerie Dixon, "Resurrecting Marcus Garvey's philosophy," *Vanguard*, June 28, 2020, https://www.vanguardngr.com/2020/06/resurrecting-marcus-garveys-philosophy1/.

2   "What the Muslims Want," excerpt from *Message to the Blackman in America*, Nation of Islam website (Accessed April 18, 2021), https://www.noi.org/muslim-program/.

3   "Marcus Garvey," *History*, November 9, 2009, https://www.history.com/topics/black-history/marcus-garvey.

4   Megan Fox, "Media Silent After White 5-Year-Old Shot Dead in Front of Family by Black Neighbor," PJ Media, August 13, 2020, https://pjmedia.com/columns/megan-fox/2020/08/13/media-silent-after-white-5-year-old-shot-dead-in-front-of-family-by-black-neighbor-n787345.

5   Benjamin Fearnow, "FBI Ranks 'Black Identity Extremists' Bigger Threat Than Al Qaeda, White Supremacists: Leaked Documents," *Newsweek*, August 8, 2019, https://www.newsweek.com/fbi-leak-black-identity-extremist-threat-1453362.

6   Bob Price and Lana Shadwick, "5 Police Officers Killed, 6 Wounded by Sniper in Dallas #BlackLivesMatter Protest," Breitbart, July 7, 2016, https://www.breitbart.com/border/2016/07/07/report-cops-shot-dallas-protest/.

7   Ben Collins, Kate Zavadski, Sarah Bertness, "Micah Johnson, Dallas Cop-Killer, Was Black Militant and Army Veteran," *Daily Beast*, July 12, 2017, https://www.thedailybeast.com/micah-johnson-dallas-cop-killer-was-black-militant-and-army-veteran.

8   Smiley N. Pool, "Five Dallas Officers Were Killed as Payback, Police Chief Says," *New York Times*, July 8, 2016, https://www.nytimes.com/2016/07/09/us/dallas-police-shooting.html.

9   Transcript of screen filmed on Project Veritas's behalf, November 6, 2019.

10 Ibid.

11 Ibid.

12 Transcript of video filmed on Project Veritas's behalf, December 11, 2019.

13 Urban Dictionary, search entry for "white trash" (Accessed May 1, 2021), https://www.urbandictionary.com/define.php?term=white%20trash.

14 Transcript of screen filmed on Project Veritas's behalf, December 6, 2019.

15 Transcript of screen filmed on Project Veritas's behalf, November 6, 2019.

16 Transcript of screen filmed on Project Veritas's behalf, September 12, 2019.

17 Monnica T. Williams, "Should people be proud of membership in a group marked by power and privilege?", *Psychology Today*, June 13, 2020, https://www.psychologytoday.com/us/blog/culturally-speaking/202006/what-is-whiteness.

18 Ibid.

19 *The Holy Bible*, King James Version, https://www.kingjamesbibleonline.org/Deuteronomy-5-9/.

20 *The Holy Bible*, King James Version, https://www.kingjamesbibleonline.org/Deuteronomy-Chapter-24/#16.

1   Transcript of screen filmed on Project Veritas's behalf, February 5, 2020.

2   Transcript of screen filmed on Project Veritas's behalf, September 12, 2019.

3   Ibid.

4   "ANTIFA: 'Practice things like an eye gouge, it takes very little pressure to injure someone's eyes'," Project Veritas, June 4, 2020, https://www.projectveritas.com/news/antifa-practice-things-like-an-eye-gouge-it-takes-very-little-pressure-to/.

5   Transcript of screen filmed on Project Veritas's behalf, September 12, 2019.

6   Bradford Betz, "Minneapolis push to defund police backfires after residents complain
    of slow response times, increase in crime," Fox News, February ,14, 2021, https://
    www.foxnews.com/us/minneapolis-defund-police-backfires-residents-complain-slow
    -response-times-increase-crime.
7   Robert Tracinski, "The DOJ's 'Trovato Report' on Ferguson," *The Federalist*, March 16,
    2015, https://thefederalist.com/2015/03/16/the-dojs-trovato-report-on-ferguson/.
8   Ryan Hartwig (@realryanhartwig), Twitter post reply to @TJMair, September 25, 2020.
    Account now deleted.
9   Ryan Hartwig, *30 Jobs in 30 Years: A Millennial's Guide to the 21st Century*, p. 123,
    Amazon, https://amzn.to/3v2mFyB.
10 Transcript of screen filmed on Project Veritas's behalf, November 17, 2019.

CHAPTER 11

1   Introductory video, Liberty Hangout the official TV Home of Kaitlin Bennett (Accessed
    April 18, 2021), https://libertyhangout.tv/join/.
2   Bryan Rolli, "Kaitlin Bennett bombarded with 'sh*t your pants' chant on college
    campus," *Daily Dot*, October 9, 2020, https://www.dailydot.com/debug/kaitlin
    -bennett-shit-pants-chant/.
3   Transcript of screen filmed on Project Veritas's behalf, February 26, 2020.
4   Ibid.
5   Ibid.
6   Ibid.
7   Ibid.
8   Transcript of screen filmed on Project Veritas's behalf, September 12, 2019.
9   Transcript of screen filmed on Project Veritas's behalf, February 26, 2020.
10 Ibid.
11 Ibid.
12 Urban Dictionary, search entry for "shit rocked" (Accessed April 18, 2021), https://www
    .urbandictionary.com/define.php?term=shitrocked.
13 Transcript of screen filmed on Project Veritas's behalf, February 26, 2020.

CHAPTER 12

1   "Tech CEOs Senate Testimony Transcript October 28," Rev.com, October 28, 2020,
    02:16:32   mark,   https://www.rev.com/blog/transcripts/tech-ceos-senate-testimony
    -transcript-october-28.
2   "Tech CEOs Senate Testimony Transcript October 28," Rev.com, October 28, 2020,
    01:25:52   mark,   https://www.rev.com/blog/transcripts/tech-ceos-senate-testimony
    -transcript-october-28.
3   Transcript of screen filmed on Project Veritas's behalf, June 6, 2019.
4   Ibid.
5   Ibid.
6   Ibid.
7   Transcript of video filmed on Project Veritas's behalf, December 6, 2019.
8   Ibid.

9   Elizabeth Dwoskin, Nitasha Tiku and Heather Kelly, "Facebook to start policing anti-Black hate speech more aggressively than anti-White comments, documents show," *Washington Post*, December 3, 2020, https://www.washingtonpost.com/technology/2020/12/03/facebook-hate-speech/.
10 "Fury as Australian senator blames Christchurch attack on Muslim immigration," *The Guardian*, March 15, 2019, https://www.theguardian.com/world/2019/mar/15/australian-senator-fraser-anning-criticised-blaming-new-zealand-attack-on-muslim-immigration.
11 "Fraser Anning punches teen after being egged while speaking to media in Melbourne," ABC News, March 15, 2019, https://www.abc.net.au/news/2019-03-16/fraser-anning-egged-in-melbourne-while-speaking-to-media/10908650.
12 Transcript of screen filmed on Project Veritas's behalf, September 12, 2019.
13 Transcript of screen filmed on Project Veritas's behalf, June 6, 2019.
14 Ibid.

CHAPTER 13

1   Transcript of screen filmed on Project Veritas's behalf, September 7, 2019.
2   Transcript of screen filmed on Project Veritas's behalf, June 6, 2019.
3   Transcript of screen filmed on Project Veritas's behalf, January 22, 2020.
4   Transcript of screen filmed on Project Veritas's behalf, June 6, 2019.
5   Ibid.
6   Transcript of screen filmed on Project Veritas's behalf, February 25th, 2020.
7   Nancy Flanders, "Facebook removes interview with abortion survivor, labels as 'spam' and 'misleading others'," Live Action, February 17, 2020, https://www.liveaction.org/news/facebook-removes-pro-lifer-interview-abortion-survivor.
8   Ibid.
9   Transcript of screen filmed on Project Veritas's behalf, September 7, 2019.
10 Transcript of screen filmed on Project Veritas's behalf, June 6, 2019.

CHAPTER 14

1   Jessica Lipsky, "Who Owns 'Boogaloo'," NPR, May 31, 2020, https://www.npr.org/sections/codeswitch/2020/05/31/864402190/who-owns-boogaloo.
2   "Boogaloo," Urban Dictionary (Accessed April 19, 2021), https://www.urbandictionary.com/define.php?term=Boogaloo.
3   Ibid.
4   Transcript of video filmed on Project Veritas's behalf, December 26, 2019.
5   Weikle for Senate (@weikleforsenate), Instagram, https://www.instagram.com/weikleforsenate/.
6   Transcript of screen filmed on Project Veritas's behalf, February 25, 2020.
7   Justin Nazaroff, email message to author, October 29, 2020.
8   Transcript of screen filmed on Project Veritas's behalf, September 12, 2019.
9   Transcript of screen filmed on Project Veritas's behalf, February 5, 2020.
10 Transcript of screen filmed on Project Veritas's behalf, January 10, 2020.
11 Ibid.

12 Jordan Davidson, "DHS Officials: NBC News Misconstrues Kyle Rittenhouse Background Information As 'Talking Points'," *The Federalist*, October 1, 2020, https://thefederalist.com/2020/10/01/dhs-officials-nbc-news-misconstrues-kyle-rittenhouse-background-information-with-talking-points/.

13 Ibid.

14 Transcript of screen filmed on Project Veritas's behalf, October 12, 2019.

15 Transcript of screen filmed on Project Veritas's behalf, June 6, 2019.

16 Ibid.

17 Ibid.

18 Transcript of screen filmed on Project Veritas's behalf, September 12, 2019.

CHAPTER 15

1 Transcript of screen filmed on Project Veritas's behalf, September 27, 2019.

2 Cathy Young, "How Facebook, Twitter silence conservative voices online," *The Hill*, October 28, 2016, https://thehill.com/blogs/pundits-blog/media/303295-how-facebook-twitter-are-systematically-silencing-conservative.

3 Reference to screen filmed on Project Veritas's behalf, October 2, 2019.

4 "Joro Olumofin apologizes to Women for his Offensive Choice of Words during Instagram Outburst," *Bella Naija*, January 27, 2016, https://www.bellanaija.com/2016/01/joro-olumofin-apologizes-to-women-for-his-offensive-choice-of-words-during-instagram-outburst/.

5 Reference to screen filmed on Project Veritas's behalf, September 24, 2019.

CHAPTER 16

1 "What is Section 230 of the Communications Decency Act," Minc Law (Accessed April 20, 2021), https://www.minclaw.com/legal-resource-center/what-is-section-230-of-the-communication-decency-act-cda/.

2 Zhanna Malekos Smith, "The Goldilocks Porridge Problem with Section 230", CSIS, November 3, 2020, https://www.csis.org/blogs/technology-policy-blog/goldilocks-porridge-problem-section-230.

3 Siladitya Ray and Rachel Sandler, "48 Attorneys General, FTC File Antitrust Charges Against Facebook," *Forbes*, December 11, 2020, https://www.forbes.com/sites/siladityaray/2020/12/09/48-us-states-ftc-file-antitrust-charges-against-facebook/?sh=29f79ed17653.

4 "FTC Sues Facebook for Illegal Monopolization," Federal Trade Commission, December 9, 2020, https://www.ftc.gov/news-events/press-releases/2020/12/ftc-sues-facebook-illegal-monopolization.

5 Jason M. Fyk, "Section 230 — 'A' Decision That Could Change 'The' Social Media World," *Medium*, October 15, 2020, https://jasonfyk.medium.com/section-230-a-decision-that-could-change-the-social-media-world-128382bf37df.

6 "Petition for a Writ of Certiorari," Jason Fyk vs Facebook Inc. (Accessed April 20, 2021), https://www.supremecourt.gov/DocketPDF/20/20-632/159405/20201102152149780_Fyk%20Petition%20E%20FILE%20Nov%202%202020.pdf.

7 Jason M. Fyk, "(FYK-S)ection 230, proposed 'proper' CDA amendment," *Medium*, April 16, 2021,https://jasonfyk.medium.com/fyk-s-ection-230-90bdad8fc10.

8 "Petition for a Writ of Certiorari."

9 Jason M. Fyk, "Section 230."

10 Ibid.

11 Jason M. Fyk, "Does Section 230 provide different immunity for Facebook rather than Malwarebytes?", *Medium*, April 12, 2021, https://jasonfyk.medium.com/does -section-230-provide-different-immunity-for-facebook-rather-than-malwarebytes -cb4415c77122.

12 Renee Diresta, "Facebook's Anti-Conservative Bias Audit Is Here," *Slate*, August 21, 2019, https://slate.com/technology/2019/08/facebook-anti-conservative-bias-audit-results -kyl.html.

13 Allum Bokhari, "The Key Points of Trump's Executive Order on Social Media Censorship," Breitbart, May 28, 2020, https://www.breitbart.com/tech/2020/05/28 /the-key-points-of-trumps-executive-order-on-social-media-censorship/.

14 Allum Bokhari, "'Break Up Big Tech Act' Sponsored by Gosar, Gabbard Takes on Social Media Censorship," Breitbart, December 10, 2020, https://www.breitbart.com /tech/2020/12/10/break-up-big-tech-act-sponsored-by-gosar-gabbard-takes-on-social -media-censorship/.

15 Ryan Hartwig, "A Look at Facebook's Sad and Severe Censorship Saga," *Medium*, July 9, 2020, https://ryanhartwig.medium.com/a-look-at-facebooks-sad-and-severe -censorship-saga-3c53cd4b8a66.

CHAPTER 17

1 Renee Diresta, "Facebook's Anti-Conservative Bias Audit Is Here," *Slate*, August 21, 2019, https://slate.com/technology/2019/08/facebook-anti-conservative-bias-audit-results -kyl.html.

2 Ibid.

3 Sean Moran, "Josh Hawley: Facebook Political Bias 'Audit' a 'Smokescreen Disguised as a Solution'," Breitbart, August 20, 2019, https://www.breitbart.com/politics/2019/08/20 /josh-hawley-facebook-political-bias-audit-a-smokescreen-disguised-as-a-solution/.

4 Transcript of video filmed on Project Veritas's behalf, November 8, 2020.

5 Jon Schweppe and Craig Parshall, "Here's A Small-Government Solution To Section 230's Big Tech Problem," *The Federalist*, February 14, 2020, https://thefederalist .com/2020/02/14/heres-a-small-government-solution-to-section-230s-big-tech -problem/.

6 "Section 230 Overview," Minc Law (Accessed April 20, 2021), https://www.minclaw.com /legal-resource-center/what-is-section-230-of-the-communication-decency-act-cda/.

CHAPTER 18

1 "Charlie Hebdo attack: Three days of terror," BBC News, January 14, 2015, https:// www.bbc.com/news/world-europe-30708237.

2 Alastair Jamieson, "'Draw Muhammad' Shooting in Garland: What We Know about Texas Attack," NBC News, May 4, 2015, https://www.nbcnews.com/news/us-news /draw-muhammad-shooting-who-was-behind-cartoon-contest-n353081.

[3] Nigel Duara, "Anti-Islam protesters stage demonstration at Phoenix mosque," *LA Times*, May 29, 2015, https://www.latimes.com/nation/la-na-anti-islam-demonstration-20150529-story.html.

[4] Transcript of screen filmed on Project Veritas's behalf, February 25, 2020.

[5] Ibid.

[6] Ibid.

[7] Ibid.

CHAPTER 19

[1] "ANOTHER Facebook Whistleblower: Interference 'On a Global Scale'," *Project Veritas*, June 25, 2020, www.projectveritas.com/news/another-facebook-whistleblower-interference-on-a-global-level-in-elections/.

[2] Ibid.

[3] Ibid.

[4] Ibid.

[5] Ibid.

[6] Ibid.

[7] Ibid.

[8] Ibid.

[9] Ibid.

[10] Ibid.

[11] Ibid.

[12] Ibid.

[13] Ibid.

[14] Ibid.

[15] Ibid.

[16] Ibid.

[17] Ibid.

[18] Ibid.

[19] Nikki Peter Petrikowski, "Charlie Hebdo Shooting," Encyclopedia Britannica (Accessed March 14, 2021), www.britannica.com/event/Charlie-Hebdo-shooting.

CHAPTER 20

[1] Casey Newton, "The Trauma Floor – The Secret Lives of Facebook Moderators in America," "The Verge," February 25, 2019, www.theverge.com/2019/2/25/18229714/cognizant-facebook-content-moderator-interviews-trauma-working-conditions-arizona.

[2] Ibid.

[3] Ibid.

[4] Ibid.

[5] Nick Hopkins, "Revealed: Facebook's Internal Rulebook on Sex, Terrorism, and Violence," *The Guardian*, May 21, 2017, www.theguardian.com/news/2017/may/21/revealed-facebook-internal-rulebook-sex-terrorism-violence.

[6] Casey Newton, "The Trauma Floor."

[7] Ibid.

8   Casey Newton, "Why a Top Content Moderation Company Quit the Business Instead of Fixing Its Problems," "The Verge," November 1, 2019, www.theverge.com /interface/2019/11/1/20941952/cognizant-content-moderation-restructuring-facebook -twitter-google.

9   Ibid.

10 Casey Newton, "Facebook will Pay $52 Million in Settlement With Moderators Who Developed PTSD on the Job," "The Verge," May 12, 2020, www.theverge .com/2020/5/12/21255870/facebook-content-moderator-settlement-scola-ptsd -mental-health.

11 Allum Bokhari, "Big Tech Whistleblowers Urge Lawmakers to Tackle Silicon Valley Bias: 'Please Do Something Already'," Breitbart, July 30, 2020, https://www.breitbart.com /tech/2020/07/30/big-tech-whistleblowers-urge-lawmakers-to-tackle-silicon-valley -bias-please-do-something-already/.

12 "Final Version Joint Letter to Congress from Three Whistleblowers" (Accessed April 20, 2021), https://www.docdroid.net/hPIv4CP/final-version-joint-letter-to-congress-from -three-whistleblowers-a-docx#page=3.

13 "The US Global Magnitsky Act," Human Rights Watch (Accessed April 20, 2021), https://www.hrw.org/news/2017/09/13/us-global-magnitsky-act.

14 "Ryan Hartwig no Programa Coliseum," YouTube, October 17, 2020, https://youtu.be /MlRR4yc1x5c.

15 "Congressman Matt Gaetz Files Criminal Referral Against Facebook CEO Mark Zuckerberg," Office of Congressman Matt Gaetz, July 27, 2020, www.gaetz.house.gov /media/press-releases/congressman-matt-gaetz-files-criminal-referral-against-facebook -ceo-mark.

16 Ibid.

17 Malwarebytes, Inc. v. Enigma Software, Supreme Court of the United States, Denial of Certiori, October 13, 2020, Statement of Justice Clarence Thomas, www.scotusblog .com/case-files/cases/malwarebytes-inc-v-enigma-software-group-usa-llc/.

CHAPTER 21

1   Peter Navarro, "The Immaculate Deception – Six Key Dimensions of Election Irregularities," December 15, 2020, www.bannonswarroom.com/wp-content/uploads /2020/12/The-Immaculate-Deception-12.15.20-1.pdf.

2   Ibid.

3   Ibid.

4   Ibid.

5   Peter Navarro, "The Art of the Steal – Volume 2 of the Navarro Report," January 5, 2021, www.4cmitv.com/wp-content/uploads/2021/01/Peter_K_Navarro_The_Art_of _the_Steal_2021JAN05.pdf.

6   Peter Navarro, "Yes, President Trump Won," January 14, 2021, www.navarroreport .com/#342e5f15-d44c-436a-bb58-9051a9397783.

7   "It's Because of Who I Am and What I Might Say – Trump Official Peter Navarro Banned From Twitter after Not Tweeting for Weeks," Newsla, January 27, 2021, www.newsla.localad.com/2021/01/27/its-because-of-who-i-am-and-what-i-might -say-trump-official-peter-navarro-banned-from-twitter-after-not-tweeting-for-weeks -video/.

8   Malcolm Gladwell, *Blink: The Power of Thinking Without Thinking*, Back Bay Books, New York (2005), pp. 32–33.

CHAPTER 22

1   Transcript of video filmed on Project Veritas's behalf, February 26, 2020.
2   Transcript of video filmed on Project Veritas's behalf, February 25, 2020.
3   Transcript of video filmed on Project Veritas's behalf, November 8, 2019.
4   Transcript of video filmed on Project Veritas's behalf, December 21, 2019.
5   Transcript of video filmed on Project Veritas's behalf, December 10, 2019.
6   Transcript of audio recorded on Project Veritas's behalf, November 19, 2019.
7   Transcript of video filmed on Project Veritas's behalf, November 20, 2019.
8   Transcript of video filmed on Project Veritas's behalf, January 25, 2020.
9   Transcript of video filmed on Project Veritas's behalf, December 20, 2019.
10   Transcript of video filmed on Project Veritas's behalf, January 18, 2020.
11   Transcript of video filmed on Project Veritas's behalf, February 4, 2020.
12   Transcript of video filmed on Project Veritas's behalf, December 11, 2019.
13   Ibid.